LONDON MATHEMATICAL SOCIETY STUDENT T

Managing editor: Professor E.B. Davies, Department of Mathematics, King's College, Strand, London WC2R 2LS

1 Introduction to combinators and -calculus, J.R. HINDLEY & J.P. SELDIN
2 Building models by games, WILFRID HODGES
3 Local fields, J.W.S. CASSELS
4 An introduction to twistor theory, S.A. HUGGETT & K.P. TOD
5 Introduction to general relativity, L. HUGHSTON & K.P. TOD
6 Lectures on stochastic analysis: diffusion theory, DANIEL W. STROOCK
7 The theory of evolution and dynamical systems, J. HOFBAUER & K. SIGMUND
8 Summing and nuclear norms in Banach space theory, G.J.O. JAMESON
9 Automorphisms of surfaces after Nielsen and Thurston, A.CASSON & S. BLEILER
10 Non-standard analysis and its applications, N.CUTLAND (ed)
11 The geometry of spacetime, G. NABER
12 Undergraduate algebraic geometry, MILES REID

London Mathematical Society Student Texts. 12

# Undergraduate Algebraic Geometry

Miles Reid, Mathematical Institute, University of Warwick

CAMBRIDGE UNIVERSITY PRESS
Cambridge
New York New Rochelle Melbourne Sydney

Published by the Press Syndicate of the University of Cambridge
The Pitt Building, Trumpington Street, Cambridge CB2 1RP
32 East 57th Street, New York, NY 10022, USA
10, Stamford Road, Oakleigh, Melbourne 3166, Australia



First published 1988

Printed in Great Britain at the University Press, Cambridge

*Library of Congress cataloging in publication data: available*

*British Library cataloguing in publication data:*
Reid, Miles
Undergraduate algebraic geometry
1. Algebraic geometry
I. Title II. Series
516.3'5

ISBN 0 521 35559 1 Hardcover
ISBN 0 521 35662 8 Paperback

# Preface

There are several good recent textbooks on algebraic geometry at the graduate level, but not (to my knowledge) any designed for an undergraduate course. Humble notes are from a course given in two successive years in the 3rd year of the Warwick undergraduate math course, and are intended as a self-contained introductory textbook.

# Contents

# §0. Woffle

This section is intended as a cultural introduction, and is not *logically* part of the course, so just skip through it.

**(0.1)** A variety is (roughly) a locus defined by polynomial equations:

$$V = \{P \in k^n \mid f_i(P) = 0\} \subset k^n,$$

where $k$ is a field and $f_i \in k[X_1, .. X_n]$ are polynomials; so for example, the plane curves $C: (f(x, y) = 0) \subset \mathbb{R}^2$ or $\mathbb{C}^2$.

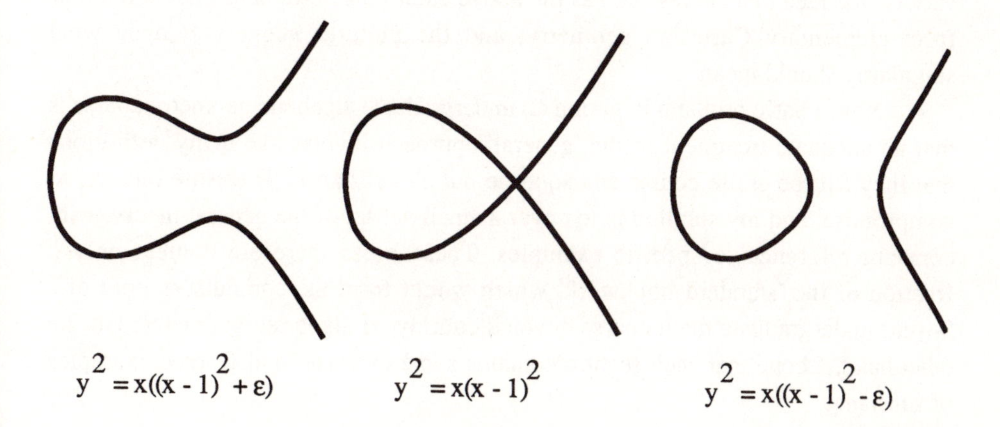

$y^2 = x((x-1)^2 + \varepsilon)$  $y^2 = x(x-1)^2$  $y^2 = x((x-1)^2 - \varepsilon)$

I want to study $V$; several questions present themselves:

**Number Theory.** For example, if $k = \mathbb{Q}$ and $V \subset \mathbb{Q}^n$, how can we tell if $V$ is non-empty, or find all its points if it is? A specific case is historically of some significance: how many solutions are there to

$$x^n + y^n = 1, \ x, y \in \mathbb{Q}, n \geq 3\ ?$$

Questions of this kind are generally known as *Diophantine problems*.

**Topology.** If $k$ is $\mathbb{R}$ or $\mathbb{C}$ (which it quite often is), what kind of topological space is $V$ ? For example, the connected components of the above cubics are obvious topological invariants.

**Singularity theory.** What kind of topological space is $V$ near $P \in V$; if $f\colon V_1 \to V_2$ is a regular map between two varieties (for example, a polynomial map $\mathbb{R}^2 \to \mathbb{R}$), what kind of topology and geometry does $f$ have near $P \in V_1$?

**(0.2)** There are two possible approaches to studying varieties:

**Particular.** Given specific polynomials $f_i$, we can often understand the variety $V$ by explicit tricks with the $f_i$; this is fun if the dimension $n$ and the degrees of the $f_i$ are small, or the $f_i$ are specially nice, but things get progressively more complicated, and there rapidly comes a time when mere ingenuity with calculations doesn't tell you much about the problem.

**General.** The study of properties of $V$ leads at once to basic notions such as regular functions on $V$, non-singularity and tangent planes, the dimension of a variety: the idea that curves such as the above cubics are 1-dimensional is familiar from elementary Cartesian geometry, and the pictures suggest at once what singularity should mean.

Now a basic problem in giving an undergraduate algebraic geometry course is that an adequate treatment of the 'general' approach involves so many definitions that they fill the entire course and squeeze out all substance. Therefore one has to compromise, and my solution is to cover a small subset of the general theory, with constant reference to specific examples. These notes therefore contain only a fraction of the 'standard bookwork' which would form the compulsory core of a 3-year undergraduate math course devoted entirely to algebraic geometry. On the other hand, I hope that each section contains some exercises and worked examples of substance.

**(0.3)** The specific flavour of algebraic geometry comes from the use of only polynomial functions (together with rational functions); to explain this, if $U \subset \mathbb{R}^2$ is an open interval, one can reasonably consider the following rings of functions on $U$:

$C^0(U)$ = all continuous functions $f\colon U \to \mathbb{R}$;

$C^\infty(U)$ = all smooth functions (that is, differentiable to any order);

$C^\omega(U)$ = all analytic functions (that is, convergent power series);

$\mathbb{R}[X]$ = the polynomial ring, viewed as polynomial functions on $U$.

There are of course inclusions $\mathbb{R}[X] \subset C^\omega(U) \subset C^\infty(U) \subset C^0(U)$.

These rings of functions correspond to some of the important categories of geometry: $C^0(U)$ to the topological category, $C^\infty(U)$ to the differentiable category (differentiable manifolds), $C^\omega$ to real analytic geometry, and $\mathbb{R}[X]$ to algebraic geometry. The point I want to make here is that each of these inclusion signs represents an absolutely *huge* gap, and that this leads to the main characteristics of geometry in the different categories. Although it's not stressed very much in school and first year university calculus, any reasonable way of measuring $C^0(U)$ will reveal that the differentiable functions have measure 0 in the continuous functions (so if you pick a continuous function at random then with probability 1 it will be nowhere differentiable, like Brownian motion). The gap between $C^\omega(U)$ and $C^\infty(U)$ is exemplified by the behaviour of $f(x) = \exp(-1/x^2)$, the standard function which is differentiable infinitely often, but for which the Taylor series (at 0) does not converge to f; using this, you can easily build a $C^\infty$ 'bump function' $f: \mathbb{R} \to \mathbb{R}$ such that $f(x) = 1$ if $|x| \leq 0.9$, and $f(x) = 0$ if $|x| \geq 1$:

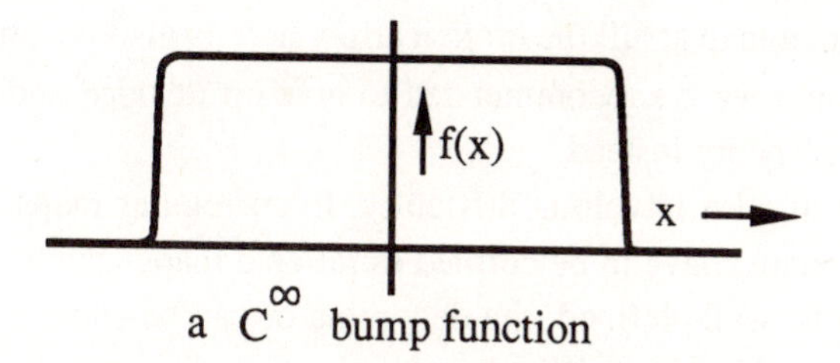


a $C^\infty$ bump function

In contrast, an analytic function on U extends (as a convergent power series) to an analytic function of a complex variable on a suitable domain in $\mathbb{C}$, so that (using results from complex analysis), if $f \in C^\omega(U)$ vanishes on a real interval, it must vanish identically. This is a kind of 'rigidity' property which characterises analytic geometry as opposed to differential topology.

**(0.4)** There are very few polynomial functions: the polynomial ring $\mathbb{R}[X]$ is just a countable-dimensional $\mathbb{R}$-vector space, whereas $C^\omega(U)$ is already uncountable. Even allowing rational functions - that is, extending $\mathbb{R}[X]$ to its field of fractions $\mathbb{R}(X)$ - doesn't help much. (2.2) will provide an example of the characteristic rigidity of the algebraic category. The fact that it is possible to construct a geometry using only this set of functions is itself quite remarkable. Not surprisingly, there are difficulties involved in setting up this theory:

**Foundations via commutative algebra.** Topology and differential topology can rely on the whole corpus of ε-δ analysis taught in a series of 1st and 2nd year undergraduate courses; to do algebraic geometry working only with polynomial rings, we need instead to study rings such as the polynomial ring $k[X_1,.. X_n]$ and its ideals. In other words, we have to develop commutative algebra in place of calculus.

The Nullstellensatz (§3 below) is a typical example of a statement having direct intuitive geometric content (essentially, 'different ideals of functions in $k[X_1,.. X_n]$ define different varieties $V \subset k^n$') whose proof involves quite a lengthy digression through finiteness conditions in commutative algebra.

**Rational maps and functions.** Another difficulty arising from the decision to work with polynomials is the necessity of introducing 'partially-defined functions'; because of the 'rigidity' hinted at above, we'll see that for some varieties (in fact for all projective varieties), there do not exist any non-constant regular functions (see Ex. 5.1, Ex. 5.12 and the discussion in (8.10)). Rational functions (that is, 'functions' of the form $f = g/h$, where $g, h$ are polynomial functions) are not defined at points where the denominator vanishes. Although reprehensible, it is a firmly entrenched tradition among algebraic geometers to use 'rational function' and 'rational map' to mean 'only partially-defined function (or map)'. So a rational map $f: V_1 \dashrightarrow V_2$ is not a map at all; the broken arrow here is also becoming traditional. Students who disapprove are recommended to give up at once and take a reading course in Category Theory instead.

This is not at all a frivolous difficulty. Even regular maps (= morphisms, these are genuine maps) have to be defined as rational maps which are regular at all points $P \in V$ (that is, well-defined, the denominator can be chosen not to vanish at P). Closely related to this is the difficulty of giving a proper intrinsic definition of a variety: in this course (and in others like it, in my experience), affine varieties $V \subset \mathbb{A}^n$ and quasiprojective varieties $V \subset \mathbb{P}^n$ will be defined, but there will be no proper definition of 'variety' without reference to an ambient space. Roughly speaking, a variety should be what you get if you glue together a number of affine varieties along isomorphic open subsets. But our present language, in which isomorphisms are themselves defined more or less explicitly in terms of rational functions, is just too cumbersome; the proper language for this glueing is sheaves, which are well treated in graduate textbooks.

**(0.5)** So much for the drawbacks of the algebraic approach to geometry. Having said this, almost all the algebraic varieties of importance in the world today are quasiprojective, and we can have quite a lot of fun with varieties without worrying overmuch about the finer points of definition.

The main advantages of algebraic geometry are that it is purely algebraically defined, and that it applies to any field, not just $\mathbb{R}$ or $\mathbb{C}$; we can do geometry over fields of characteristic p. (Don't say 'characteristic p – big deal, that's just the finite fields'; to start with, very substantial parts of group theory are based on geometry over finite fields, as are large parts of combinatorics used in computer science. Next, there are lots of interesting fields of characteristic p other than finite ones. Moreover, at a deep level, the finite fields are present and working inside $\mathbb{Q}$ and $\mathbb{C}$. Most of the deep results on arithmetic of varieties over $\mathbb{Q}$ use a considerable amount of geometry over $\mathbb{C}$ or over the finite fields and their algebraic closures.)

This concludes the introduction; see the informal discussion in (2.15) and the final §8 for more general culture.

**(0.6)** As to the structure of the book, Chapter I and Chapter III aim to indicate some worthwhile problems which can be studied by means of algebraic geometry. Chapter II is an introduction to the commutative algebra referred to in (0.4) and to the categorical framework of algebraic geometry; the student who is prone to headaches could perhaps take some of the proofs for granted here, since the material is standard, and the author is a professional algebraic geometer of the highest moral fibre.

§8 contains odds and ends that may be of interest or of use to the student, but that don't fit in the main text: a little of the history and sociology of the modern subject, hints as to relations of the subject-matter with more advanced topics, technical footnotes, etc.

## Course Prerequisites:

**Algebra:** Quadratic forms, easy properties of commutative rings and their ideals, principal ideal domains and unique factorisation.

**Galois Theory:** Fields, polynomial rings, finite extensions, algebraic versus transcendental extensions, separability.

**Topology and geometry:** Definition of topological space, projective space $\mathbb{P}^n$ (but I'll go through it again in detail).

**Calculus in $\mathbb{R}^n$:** Partial derivatives, implicit function theorem (but I'll remind you of what I need when we get there).

**Commutative algebra:** Other experience with commutative rings is desirable, but not essential.

## Course relates to:

**Complex Function Theory.** An algebraic curve over $\mathbb{C}$ is a 1-dimensional complex manifold, and regular functions on it are holomorphic, so that this course is closely related to complex function theory, even if the relation is not immediately apparent.

**Algebraic Number Theory.** For example the relation with Fermat's Last Theorem.

**Catastrophe Theory.** Catastrophes are singularities, and are essentially always given by polynomial functions, so that the analysis of the geometry of the singularities is pure algebraic geometry.

**Commutative Algebra.** Algebraic geometry provides motivation for commutative algebra, and commutative algebra provides technical support for algebraic geometry, so that the two subjects enrich one another.

### Exercises to §0.

**0.1.** (a) Show that for fixed values of $(y, z)$, $x$ is a repeated root of $x^3 + zx - y = 0$ if and only if $x = 3y/2z$ and $27y^2 + 4z^3 = 0$;

(b) there are 3 distinct roots if and only if $27y^2 + 4z^3 < 0$;

(c) sketch the surface $S: (x^3 + zx - y = 0)$ in $\mathbb{R}^3$ and its projection onto the $(y, z)$-plane;

(d) now open up any book or article on catastrophe theory and compare.

**0.2.** Let $f \in \mathbb{R}[X, Y]$ and let $C: (f = 0) \subset \mathbb{R}^2$; say that $P \in C$ is isolated if there is an $\varepsilon > 0$ such that $C \cap B(P, \varepsilon) = P$. Show by example that $C$ can have isolated points.

Prove that if $P \in C$ is an isolated point then $f: \mathbb{R}^2 \to \mathbb{R}$ must have a max or min at $P$, and deduce that $\partial f/\partial x$ and $\partial f/\partial y$ vanish at $P$. This proves that an isolated point of a real curve is singular.

**0.3.** **Cubic curves:** (i) Draw the graph of $f(x) = 4x^3 + 6x^2$ and its intersection with the horizontal lines $y = t$ for integer values of $t \in [-1, 5]$; (ii) draw the cubic curves $y^2 = 4x^3 + 6x^2 + t$ for the same values of $t$.

## Books

Most of the following are textbooks at a graduate level, and some are referred to in the text:

**W. Fulton**, Algebraic curves, Springer. (This is the most down-to-earth and self-contained of the graduate texts; Chaps.1-6 are quite well suited to an undergraduate course, although the material is somewhat dry.)

**I.R. Shafarevich**, Basic algebraic geometry, Springer. (A graduate text, but Ch. I, and §II.1 are quite suitable material.)

**P. Griffiths and J. Harris**, Principles of algebraic geometry, Wiley. (Gives the complex analytic point of view.)

**D. Mumford**, Algebraic geometry I, Complex projective varieties, Springer.

**D. Mumford**, Introduction to algebraic geometry, Harvard notes. (Not immediately very readable, but goes directly to the main points; many algebraic geometers of my generation learned their trade from these notes. It can be found (little red book) in many university libraries.)

**K. Kendig**, Elementary algebraic geometry, Springer. (Treats the relation between algebraic geometry and complex analytic geometry.)

**R. Hartshorne**, Algebraic geometry, Springer. (This is the professional's handbook, and covers much more advanced material; Ch. I is an undergraduate course in bare outline.)

**M. Berger**, Geometry I and II, Springer. (Some of the material of the sections on quadratic forms and quadric hypersurfaces in II is especially relevant.)

**M.F. Atiyah and I.G. Macdonald**, Commutative algebra, Addison-Wesley. (An invaluable textbook.)

**E. Kunz**, Introduction to commutative algebra and algebraic geometry, Birkhäuser.

**H. Matsumura**, Commutative ring theory, Cambridge. (A more detailed text on commutative algebra.)

**D. Mumford**, Curves and their Jacobians, Univ. of Michigan Press. (Colloquial lectures, going quite deep quite fast.)

**C.H. Clemens**, A scrapbook of complex curves, Plenum. (Lots of fun.).

**E. Brieskorn and H. Knörrer**, Plane algebraic curves, Birkhäuser.

**A. Beauville**, Complex algebraic surfaces, LMS Lecture Notes, Cambridge.

**J. Kollár**, The structure of algebraic threefolds: An introduction to Mori's program, Bull. Amer. Math. Soc. **17** (1987), 211-273. (A nicely presented travel brochure to one active area of research. Mostly harmless.)

**J.G. Semple and L. Roth**, Introduction to algebraic geometry, Oxford. (A marvellous old book, full of information, but almost entirely lacking in rigour.)

**J.L. Coolidge**, Treatise on algebraic plane curves, Oxford and Dover.

# Chapter I. Playing with plane curves

## §1. Plane conics

I start by studying the geometry of conics as motivation for the projective plane $\mathbb{P}^2$. Projective geometry is usually mentioned in 2nd year undergraduate geometry courses, and I recall some of the salient features, with some emphasis on homogeneous coordinates, although I completely ignore the geometry of linear subspaces and the 'cross-ratio'. The most important aim for the student should be to grasp the way in which geometrical ideas (for example, the idea that 'points at infinity' correspond to asymptotic directions of curves) are expressed in terms of coordinates. The interplay between the intuitive geometrical picture (which tells you what you should be expecting), and the precise formulation in terms of coordinates (which allows you to cash in on your intuition) is a fascinating aspect of algebraic geometry.

**(1.1) Example of a parametrised curve.** Pythagoras' Theorem says that, in the diagram

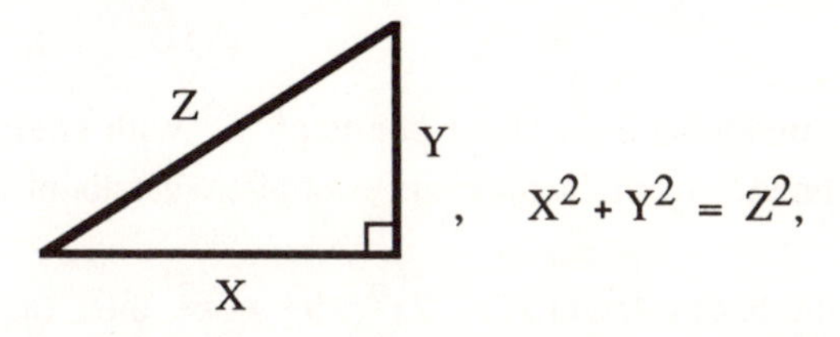


so (3, 4, 5) and (5, 12, 13), as every ancient Egyptian knew. How do you find all integer solutions? The equation is homogeneous, so that $x = X/Z$, $y = Y/Z$ gives the circle $C : (x^2 + y^2 = 1) \subset \mathbb{R}^2$, which can easily be seen to be parametrised as

$$x = 2\lambda/(\lambda^2 + 1),\ \ y = (\lambda^2 - 1)/(\lambda^2 + 1), \quad \text{where} \quad \lambda = x/(1 - y);$$

so this gives all solutions:

$$X = 2\ell m,\ \ Y = \ell^2 - m^2,\ \ Z = \ell^2 + m^2 \quad \text{with} \quad \ell, m \in \mathbb{Z} \text{ coprime},$$

(or each divided by 2 if $\ell$, m are both odd). Note that the equation is homogeneous, so that if (X, Y, Z) is a solution, then so is $(\lambda X, \lambda Y, \lambda Z)$.

Maybe the parametrisation was already familiar from school geometry, and in any case, it's easy to verify that it works. However, if I didn't know it already, I could have obtained it by an easy geometrical argument, namely linear projection from a given point:

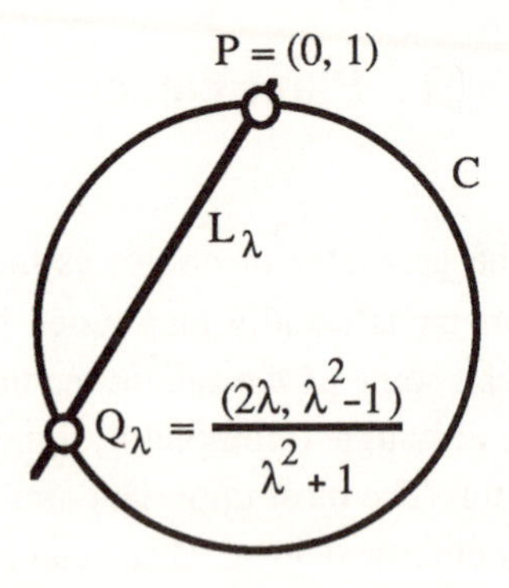


$P = (0, 1) \in C$, and if $\lambda \in \mathbb{Q}$ is any value, then the line $L_\lambda$ through $P$ with slope $-\lambda$ meets $C$ in a further point $Q_\lambda$. This construction of a map by means of linear projection will appear many times in what follows.

**(1.2) Similar example.** $C$: $(2X^2 + Y^2 = 5Z^2)$. The same method leads to the parametrisation $\mathbb{R} \to C$ given by

$$x = \frac{2\sqrt{5}\lambda}{1 + 2\lambda^2}, \quad y = \frac{2\lambda^2 - 1}{1 + 2\lambda^2}.$$

This allows us to understand all about points of $C$ with coefficients in $\mathbb{R}$, and there's no real difference from the previous example; what about $\mathbb{Q}$?

**Proposition.** If $(a, b, c) \in \mathbb{Q}$ satisfies $2a^2 + b^2 = 5c^2$ then $(a, b, c) = (0, 0, 0)$.

**Proof.** Multiplying through by a common denominator and taking out a common factor if necessary, I can assume that $a, b, c$ are integers, not all of which are divisible by 5; also if $5 \mid a$ and $5 \mid b$ then $25 \mid 5c^2$, so that $5 \mid c$, which contradicts what I've just said. It is now easy to get a contradiction by considering the possible values of $a$ and $b$ mod 5: since any square is 0, 1 or 4 mod 5, clearly $2a^2 + b^2$ is one of $0+1, 0+4, 2+0, 2+1, 2+4, 8+0, 8+1$ or $8+4$ mod 5, none of which can be of the form $5c^2$. Q.E.D.

Note that this is a thoroughly arithmetic argument.

**(1.3) Conics in $\mathbb{R}^2$.** A conic in $\mathbb{R}^2$ is a plane curve given by a quadratic equation

$$q(x,y) = ax^2 + bxy + cy^2 + dx + ey + f = 0.$$

Everyone has seen the classification of non-degenerate conics:

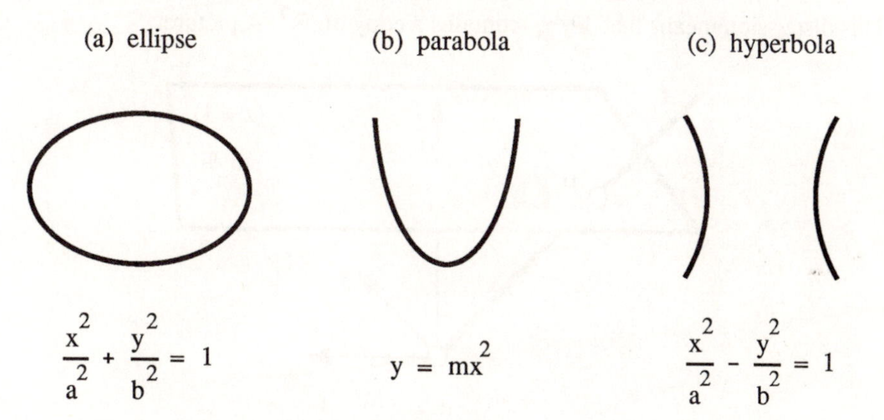


in addition, there are a number of peculiar cases:

(d) single point given by $x^2 + y^2 = 0$;

(e, f, g) empty set given by any of the 3 equations: (e) $x^2 + y^2 = -1$, (f) $x^2 = -1$ or (g) $0 = 1$. These three equations are different, although they define the same locus of zeros in $\mathbb{R}^2$; consider for example their complex solutions.

(h) line $x = 0$;

(i) line pair $xy = 0$;

(j) parallel lines $x(x - 1) = 0$;

(k) 'double line' $x^2 = 0$;

you can choose for yourself whether you'll allow the final case:

(l) whole plane given by $0 = 0$.

**(1.4) Projective plane.** The definition 'out of the blue':

$$\begin{aligned} \mathbb{P}^2_{\mathbb{R}} &= \{\text{lines of } \mathbb{R}^3 \text{ through origin}\} \\ &= \{\text{ratios } X : Y : Z\} \\ &= (\mathbb{R}^3 \setminus \{0\})/\sim, \text{ where } (X, Y, Z) \sim (\lambda X, \lambda Y, \lambda Z) \text{ if } \lambda \in \mathbb{R} \setminus \{0\}. \end{aligned}$$

(The sophisticated reader will have no difficulty in generalising from $\mathbb{R}^3$ to an arbitrary vector space over a field, and in replacing work in a chosen coordinate system with intrinsic arguments.)

To represent a ratio $X:Y:Z$ for which $Z \neq 0$, I can set $x = X/Z$, $y = Y/Z$; this simplifies things, since the ratio corresponds to just two real numbers. In other words, the equivalence class of $(X, Y, Z)$ under $\sim$ has a unique representative $(x, y, 1)$ with 3rd coordinate $= 1$. Unfortunately, sometimes $Z$ might be $= 0$, so that this way of choosing a representative of the equivalence class is then no good. This discussion means that $\mathbb{P}^2_{\mathbb{R}}$ contains a copy of $\mathbb{R}^2$. A picture:

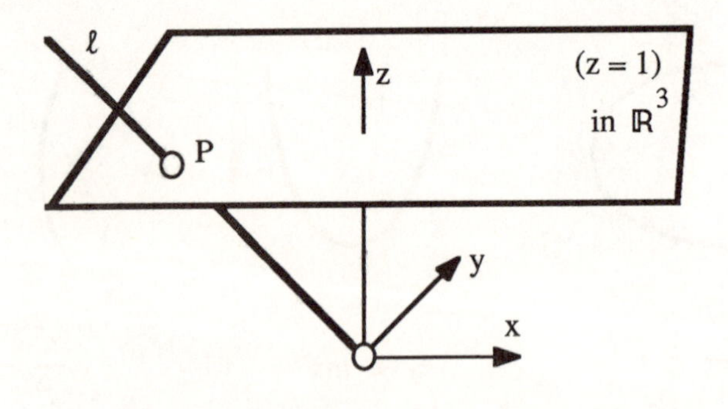


$$\mathbb{R}^2 \hookrightarrow \mathbb{R}^3 \setminus \{0\} \to \mathbb{P}^2_{\mathbb{R}} \text{ by } (x, y) \mapsto (x, y, 1)$$

the general line in $\mathbb{R}^3$ through $0$ is not contained in the plane $(Z = 0)$, so that it meets $(Z = 1)$ in exactly one point, which is a representative for that equivalence class. The lines in $(Z = 0)$ never meet $(Z = 1)$, so they correspond not to points of $\mathbb{R}^2$, but to *asymptotic directions* , or to pencils of parallel lines of $\mathbb{R}^2$; so you can think of $\mathbb{P}^2_{\mathbb{R}}$ as consisting of $\mathbb{R}^2$ together with one 'point at infinity' for every pencil of parallel lines. From this point of view, you calculate in $\mathbb{R}^2$, try to guess what's going on at infinity by some kind of 'asymptotic' argument, then (if necessary), prove it in terms of homogeneous coordinates. The definition in terms of lines in $\mathbb{R}^3$ makes this respectable, since it treats all points of $\mathbb{P}^2_{\mathbb{R}}$ on an equal footing.

Groups of transformations are of central importance throughout geometry; properties of a geometrical figure must be invariant under the appropriate kind of transformations before they are significant. An *affine* change of coordinates in $\mathbb{R}^2$ is of the form $T(\mathbf{x}) = A\mathbf{x} + B$, where $\mathbf{x} = (x, y) \in \mathbb{R}^2$, and $A$ is a $2 \times 2$ invertible matrix, $B$ a translation vector; if $A$ is orthogonal then the transformation $T$ is *Euclidean* . As everyone knows, every non-degenerate conic can be reduced to one of the standard forms (a–c) above by a Euclidean transformation. It is an exercise to the reader to show that every conic can be reduced to one of the forms (a–l) by an affine transformation.

A *projectivity*, or projective transformation of $\mathbb{P}^2_{\mathbb{R}}$ is of the form

$T(\mathbf{X}) = M\mathbf{X}$, where $M$ is an invertible $3 \times 3$ matrix. It's easy to understand the effect of this transformation on the affine piece $\mathbb{R}^2 \subset \mathbb{P}^2_{\mathbb{R}}$ : as a partially defined map $\mathbb{R}^2 \dashrightarrow \mathbb{R}^2$, it is the fractional-linear transformation

$$\begin{bmatrix} x \\ y \end{bmatrix} \mapsto \left(A\begin{bmatrix} x \\ y \end{bmatrix} + B\right)/(cx + dy + e),$$

where

$$M = \left[\begin{array}{cc|c} \multicolumn{2}{c|}{A} & B \\ \hline c & d & e \end{array}\right].$$

$T$ is of course not defined when $cx + dy + e = 0$. Perhaps this looks rather unintuitive, but it really occurs in nature: two different photographs of the same (plane) object are obviously related by a projectivity; see for example [Berger, 4.7.4] for pictures. So a math graduate getting a job interpreting satellite photography (whether for the peaceful purposes of the Forestry Commission, or as part of the vast career prospects opened up by President Reagan's defence policy) will spend a good part of his or her time computing projectivities.

Projective transformations are implicitly in use throughout these notes, usually in the form 'by a suitable choice of coordinates, I can assume ..'.

**(1.5) Equation of a conic.** The inhomogeneous quadratic polynomial

$$q(x,y) = ax^2 + bxy + cy^2 + dx + ey + f$$

corresponds to the homogeneous quadratic

$$Q(X, Y, Z) = aX^2 + bXY + cY^2 + dXZ + eYZ + fZ^2;$$

the correspondence is easy to understand as a recipe, or you can think of it as the bijection $q \longleftrightarrow Q$ given by

$$q(x, y) = Q(X/Z, Y/Z) \quad \text{with} \quad x = X/Z,\ y = Y/Z$$

and inversely,

$$Q = Z^2 q(X/Z, Y/Z).$$

A *conic* $C \subset \mathbb{P}^2$ is the curve given by $C\colon (Q(X, Y, Z) = 0)$, where $Q$ is a homogeneous quadratic expression; note that the condition $Q(X, Y, Z) = 0$ is well-defined on the equivalence class, since $Q(\lambda\mathbf{X}) = \lambda^2 Q(\mathbf{X})$ for any $\lambda \in \mathbb{R}$. As an exercise, check that the projective curve $C$ meets the affine piece $\mathbb{R}^2$ in the affine conic given by $(q = 0)$.

**'Line at infinity' and asymptotic directions.** Points of $\mathbb{P}^2$ with $Z = 0$ correspond to ratios $(X : Y : 0)$. These points form the 'line at infinity', a copy of $\mathbb{P}^1_{\mathbb{R}} = \mathbb{R} \cup \{\infty\}$ (since $(X : Y) \mapsto X/Y$ defines a bijection $\mathbb{P}^1_{\mathbb{R}} \to \mathbb{R} \cup \{\infty\}$).

A line in $\mathbb{P}^2$ is by definition given by $L: (aX + bY + cZ = 0)$, and

$$L \text{ passes through } (X, Y, 0) \iff aX + bY = 0.$$

In affine coordinates the same line is given by $ax + by + c = 0$, so that all lines with the same ratio $a : b$ pass through the same point at infinity. This is called 'parallel lines meet at infinity'.

**Examples.** (a) The hyperbola $(x^2/a^2 - y^2/b^2 = 1)$ in $\mathbb{R}^2$ corresponds to $C: (X^2/a^2 - Y^2/b^2 = Z^2)$ in $\mathbb{P}^2_{\mathbb{R}}$; clearly this meets $(Z = 0)$ in the two points $(b, \pm a, 0) \in \mathbb{P}^2_{\mathbb{R}}$, corresponding in the obvious way to the asymptotic lines of the hyperbola.

Note that in the affine piece $(X \neq 0)$ of $\mathbb{P}^2_{\mathbb{R}}$, the affine coordinates are $u = Y/X$, $v = Z/X$, so that $C$ becomes the ellipse $(u^2/b^2 + v^2 = 1/a^2)$. See Ex. 1.7 for an artistic interpretation.

(b) The parabola $(y = mx^2)$ in $\mathbb{R}^2$ corresponds to $C: (YZ = mX^2)$ in $\mathbb{P}^2_{\mathbb{R}}$; this now meets $(Z = 0)$ at the single point $(0, 1, 0)$. So in $\mathbb{P}^2$, the 'two branches of the parabola meet at infinity'; note that this is a statement with intuitive content (maybe you feel it's pretty implausible?), but is not a result you could arrive at just by contemplating within $\mathbb{R}^2$ – maybe it's not even meaningful.

**(1.6) Classification of conics in $\mathbb{P}^2$.** Let $k$ be any field of characteristic $\neq 2$; recall two results from the linear algebra of quadratic forms:

**Proposition (A).** There are natural bijections

$$\left\{\begin{matrix}\text{homogeneous}\\ \text{quadratic polys}\end{matrix}\right\} = \left\{\begin{matrix}\text{quad. forms}\\ k^3 \to k\end{matrix}\right\} \xleftrightarrow{\text{bij}} \left\{\begin{matrix}\text{symmetric bilinear}\\ \text{forms on } k^3\end{matrix}\right\},$$

given in formulas by

$$aX^2 + 2bXY + cY^2 + 2dXZ + 2eYZ + fZ^2 \longleftrightarrow \begin{bmatrix} a & b & d \\ b & c & e \\ d & e & f \end{bmatrix}.$$

A quadratic form is *non-degenerate* if the corresponding bilinear form is non-degenerate, that is, its matrix is non-singular.

**Theorem (B).** Let V be a vector space over k and $Q\colon V \to k$ a quadratic form; then there exists a basis of V such that

$$Q = \varepsilon_1 x_1^2 + \varepsilon_2 x_2^2 + .. \varepsilon_n x_n^2,$$

with $\varepsilon_i \in k$.

(This is proved by *Gram-Schmidt orthogonalisation*, if that rings a bell.) Obviously, for $\lambda \in k \smallsetminus \{0\}$ the substitution $x_i \mapsto \lambda x_i$ takes $\varepsilon_i$ into $\lambda^{-2}\varepsilon_i$.

**Corollary**. In a suitable system of coordinates, any conic in $\mathbb{P}^2_{\mathbb{R}}$ is one of the following:

(α) non-degenerate conic, C: $(X^2 + Y^2 - Z^2 = 0)$;

(β) empty set, given by $(X^2 + Y^2 + Z^2 = 0)$;

(γ) line pair, given by $(X^2 - Y^2 = 0)$;

(δ) one point (0, 0, 1), given by $(X^2 + Y^2 = 0)$;

(ε) double line, given by $(X^2 = 0)$.

(Optionally you have the whole of $\mathbb{P}^2_{\mathbb{R}}$ given by $(0 = 0)$.)

**Proof**. Any real number ε is either 0, or ± a square, so that I only have to consider Q as in the theorem with $\varepsilon_i = 0, \pm 1$. In addition, since I'm only interested in the locus $(Q = 0)$, I'm allowed to multiply Q through by $-1$. This leads at once to the given list. Q.E.D.

There are two points to make about this corollary: firstly, the list is quite a lot shorter than that in (1.3); for example, the 3 non-degenerate cases (ellipse, parabola, hyperbola) of (1.3) all correspond to case (α), and the 2 cases of intersecting and parallel line pairs are not distinguished in the projective case. Secondly, the derivation of the list from general algebraic principles is much simpler.

**(1.7) Parametrisation of a conic.** Let C be a non-degenerate, non-empty conic of $\mathbb{P}^2_{\mathbb{R}}$. Then by Corollary 1.6, taking new coordinates (X+Y, X−Y, Z), C is projectively equivalent to the curve $(XZ = Y^2)$; this is the curve parametrised by

$$\Phi\colon \mathbb{P}^1_{\mathbb{R}} \longrightarrow C \subset \mathbb{P}^2_{\mathbb{R}},$$

$$(U:V) \longmapsto (U^2 : UV : V^2).$$

**Remarks 1.** The inverse map $\Psi\colon C \to \mathbb{P}^1_{\mathbb{R}}$ is given by $(X{:}Y{:}Z) \mapsto (X{:}Y) = (Y{:}Z)$; here the left-hand ratio is defined if $X \neq 0$, and the right-hand ratio if $Z \neq 0$. In terminology to be introduced later, $\Phi$ and $\Psi$ are inverse isomorphisms of varieties.

**2.** The same thing holds for any non-empty, non-degenerate conic over a field $k$ of characteristic $\neq 2$ (see Ex. 1.5).

**(1.8) Homogeneous form in 2 variables.** Let $F(U, V)$ be a non-zero homogeneous polynomial of degree $d$ in $U, V$, with coefficients in a fixed field $k$; (I will follow tradition, and use the word *form* for 'homogeneous polynomial'):

$$F(U,V) = a_dU^d + a_{d-1}U^{d-1}V + .. a_iU^iV^{d-i} + .. a_0V^d.$$

$F$ has an associated inhomogeneous polynomial in 1 variable,

$$f(u) = a_du^d + a_{d-1}u^{d-1} + .. a_iu^i + .. a_0.$$

Clearly for $\alpha \in k$,

$$f(\alpha) = 0 \iff (u - \alpha) \mid f(u) \iff (U - \alpha V) \mid F(U, V) \iff F(\alpha, 1) = 0;$$

so zeros of $f$ correspond to zeros of $F$ on $\mathbb{P}^1$ away from the point $(1, 0)$, the 'point $\alpha = \infty$'. What does it mean for $F$ to have a zero at infinity?

$$F(1, 0) = 0 \iff a_d = 0 \iff \deg f < d.$$

Now define the *multiplicity* of a zero of $F$ on $\mathbb{P}^1$ to be

(i) the multiplicity of $f$ at the corresponding $\alpha \in k$;

or (ii) $d - \deg f$ if $(1, 0)$ is the zero.

So the multiplicity of zero of $F$ at a point $(\alpha, 1)$ is the greatest power of $(U - \alpha V)$ dividing $F$, and at $(1, 0)$ it is the greatest power of $V$ dividing $F$.

**Proposition.** Let $F(U, V)$ be a non-zero form of degree $d$ in $(U, V)$. Then $F$ has at most $d$ zeros on $\mathbb{P}^1$; furthermore, if $k$ is algebraically closed, then $F$ has exactly $d$ zeros on $\mathbb{P}^1$ provided these are counted with multiplicities as defined above.

**Proof.** Let $m_\infty$ be the multiplicity of the zero of $F$ at $(1, 0)$; then by definition, $d - m_\infty$ is the degreee of the inhomogeneous polynomial $f$, and the proposition reduces to the well-known fact that a polynomial in one variable has at most $\deg f$ roots. Q.E.D.

Note that over an algebraically closed field, $F$ will factorise as a product $F = \Pi \lambda_i^{m_i}$ of linear forms $\lambda_i = (a_iU + b_iV)$, and treated in this way, the point $(1, 0)$ corresponds to the form $\lambda_\infty = V$, and is on the same footing as all other points.

**(1.9) Easy cases of Bézout's Theorem.** Bézout's theorem say that if $C$ and $D$ are plane curves of degree $\deg C = m$, $\deg D = n$, then the number of points of intersection of $C$ and $D$ is $mn$, provided that (i) the field is algebraically closed; (ii) points of intersection are counted with the right multiplicities; (iii) we work in $\mathbb{P}^2$ to take right account of intersections 'at infinity'. See for example [Fulton, p.112] for a self-contained proof. In this section I am going to treat the case when one of the curves is a line or conic.

**Theorem.** Let $L \subset \mathbb{P}^2_k$ be a line (respectively $C \subset \mathbb{P}^2_k$ a non-degenerate conic), and let $D \subset \mathbb{P}^2_k$ be a curve defined by $D : (G_d(X, Y, Z) = 0)$, where $G$ is a form of degree $d$ in $X, Y, Z$. Assume that $L \not\subset D$ (respectively, $C \not\subset D$); then

$$\#\{L \cap D\} \leq d \qquad \text{(respectively } \#\{C \cap D\} \leq 2d).$$

In fact there is a natural definition of multiplicity of intersection such that the inequality still hold for 'number of points counted with multiplicities', and if $k$ is algebraically closed then equality holds.

**Proof.** A line $L \subset \mathbb{P}^2_k$ is given by an equation $\lambda = 0$, with $\lambda$ a linear form; for my purpose, it is convenient to give it parametrically as

$$X = a(U, V), \quad Y = b(U, V), \quad Z = c(U, V),$$

where $a, b, c$ are linear forms in $U, V$. So for example, if $\lambda = \alpha X + \beta Y + \gamma Z$, and $\gamma \neq 0$, then $L$ can be given as

$$X = U, \quad Y = V, \quad Z = -(\alpha/\gamma)U - (\beta/\gamma)V.$$

Similarly, using Corollary 1.6, a non-degenerate conic can be given parametrically as

$$X = a(U, V), \quad Y = b(U, V), \quad Z = c(U, V),$$

where $a, b, c$ are quadratic forms in $U, V$. This is because $C$ is a projective transformation of $(XZ = Y^2)$, which is parametrically $(X, Y, Z) = (U^2, UV, V^2)$, so $C$ is given by

$$\begin{bmatrix} X \\ Y \\ Z \end{bmatrix} = M \begin{bmatrix} U^2 \\ UV \\ V^2 \end{bmatrix},$$

where $M$ is a non-singular $3 \times 3$ matrix.

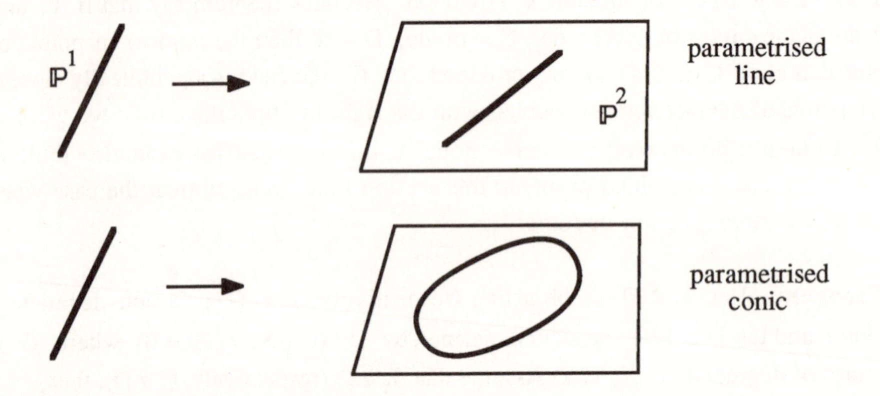


Then the intersection of $L$ (respectively $C$) with $D$ is given by finding the values of the ratios $(U:V)$ such that

$$F(U, V) = G_d(a(U,V), b(U,V), c(U,V)) = 0.$$

But $G$ is a form of degree $d$ (respectively $2d$) in $U, V$, so the result follows by (1.8).

**(1.10) Corollary.** If $P_1,.. P_5 \in \mathbb{P}^2_{\mathbb{R}}$ are distinct points such that no 4 are collinear, then there exists at most one conic through $P_1,.. P_5$.

**Proof.** Suppose by contradiction that $C_1$ and $C_2$ are conics with $C_1 \neq C_2$ such that

$$C_1 \cap C_2 \supset \{P_1,.. P_5\}.$$

$C_1$ is non-empty, so that if it's non-degenerate, then by (1.6), it's projectively equivalent to the parametrised curve

$$C_1 = \{(U^2, UV, V^2) \mid (U, V) \in \mathbb{P}^1\};$$

by (1.9), $C_1 \subset C_2$. Now if $Q_2$ is the equation of $C_2$, then $Q_2(U^2, UV, V^2) \equiv 0$ for all $(U, V) \in \mathbb{P}^1$, and an easy calculation (see Ex. 1.6) shows that $Q_2$ is a multiple of $(XZ - Y^2)$; this contradicts $C_1 \neq C_2$.

Now suppose $C_1$ is degenerate; by (1.6) again, it's either a line pair or a line, and one sees easily that

$$C_1 = L_0 \cup L_1, \quad C_2 = L_0 \cup L_2,$$

with $L_1, L_2$ distinct lines. Then $C_1 \cap C_2 = L_0 \cup (L_1 \cap L_2)$:

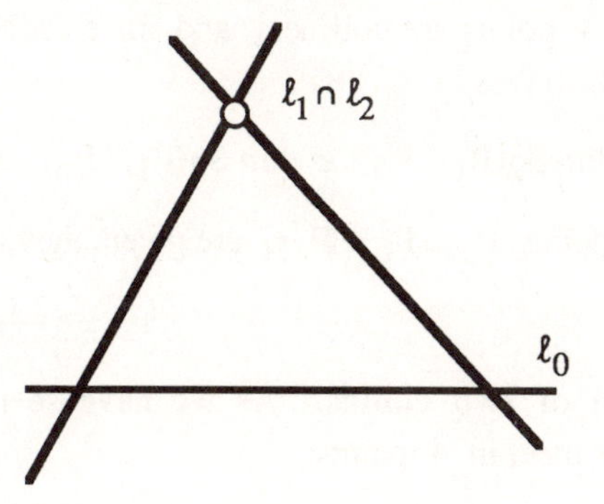


thus 4 points out of $P_1, \ldots P_5$ lie on $L_0$, a contradiction. Q.E.D.

**(1.11) Space of all conics.** Let

$$S_2 = \{\text{quadratic forms on } \mathbb{R}^3\} = \{3 \times 3 \text{ symmetric matrixes}\} \cong \mathbb{R}^6.$$

If $Q \in S_2$, write $Q = aX^2 + 2bXY + \ldots fZ^2$; then for $P_0 = (X_0, Y_0, Z_0) \in \mathbb{P}^2_{\mathbb{R}}$, consider the relation $P_0 \in C : (Q = 0)$. This is of the form

$$Q(X_0, Y_0, Z_0) = aX_0^2 + 2bX_0Y_0 + \ldots fZ_0^2 = 0,$$

and for fixed $P_0$, this is a linear equation in $(a, b, \ldots f)$. So

$$S_2(P_0) = \{Q \in S_2 \mid Q(P_0) = 0\} \cong \mathbb{R}^5 \subset S_2 = \mathbb{R}^6$$

is a 5-dimensional hyperplane. For $P_1, \ldots P_n \in \mathbb{P}^2_{\mathbb{R}}$, define similarly

$$S_2(P_1, \ldots P_n) = \{Q \in S_2 \mid Q(P_i) = 0 \text{ for } i = 1, \ldots n\};$$

then there are $n$ linear equations in the 6 coefficients $(a, b, \ldots f)$ of $Q$. This gives the result:

**Proposition.** $\dim S_2(P_1, \ldots P_n) \geq 6 - n.$

We can also expect that 'equality holds if $P_1,\dots P_n$ are general enough'. More-precisely:

**Corollary.** If $n \leq 5$ and no 4 of $P_1,\dots P_n$ are collinear, then

$$\dim S_2(P_1,\dots P_n) = 6 - n.$$

**Proof.** Corollary 1.10 implies that if $n = 5$, $\dim S_2(P_1,\dots P_5) \leq 1$, which gives the corollary in this case. If $n \leq 4$, then I can add in points $P_{n+1},\dots P_5$ while preserving the condition that no 4 points are collinear, and since each point imposes at most one linear condition, this gives

$$1 = \dim S_2(P_1,\dots P_5) \geq \dim S_2(P_1,\dots P_n) - (n - 5).$$

Note that if 6 points $P_1,\dots P_6 \in \mathbb{P}^2_{\mathbb{R}}$ are given, they may or may not lie on a conic. Q.E.D.

**(1.12) Intersection of two conics.** As we have seen above, it will often happen that two conics meet in 4 points:

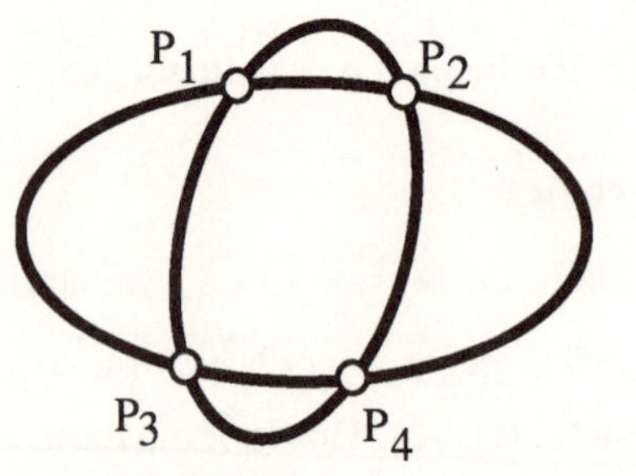


conversely according to (1.11), given 4 points $P_1,\dots P_4 \in \mathbb{P}^2$, under suitable conditions $S_2(P_1,\dots P_4)$ is a 2-dimensional vector space, so choosing a basis $Q_1, Q_2$ for $S_2(P_1,\dots P_4)$ gives 2 conics $C_1, C_2$ such that $C_1 \cap C_2 = \{P_1,\dots P_4\}$. There are lots of possibilities for degenerate intersections:

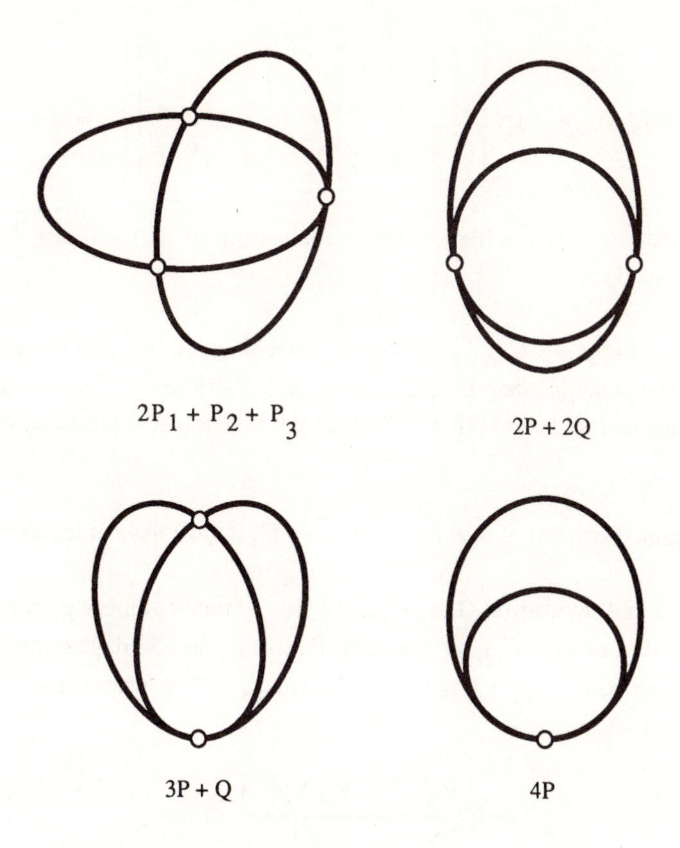


see Ex. 1.9 for suitable equations.

**(1.13) Degenerate conics in a pencil.**
**Definition.** A *pencil of conics* is a family of the form

$$C_{(\lambda,\mu)} : (\lambda Q_1 + \mu Q_2 = 0);$$

each element is a plane curve, depending in a linear way on the parameters $(\lambda, \mu)$; think of the ratio $(\lambda : \mu)$ as a point of $\mathbb{P}^1$.

Looking at the examples, one expects that for special values of $(\lambda : \mu)$ the conic $C_{(\lambda,\mu)}$ is degenerate. In fact, writing $\det(Q)$ for the determinant of the symmetric $3 \times 3$ matrix corresponding to the quadratic form $Q$, it is clear that

$$C_{(\lambda, \mu)} \text{ is degenerate} \iff \det(\lambda Q_1 + \mu Q_2) = 0.$$

Writing out $Q_1$ and $Q_2$ as symmetric matrixes expresses this condition as

$$F(\lambda, \mu) = \det\left|\lambda\begin{bmatrix} a & b & d \\ b & c & e \\ d & e & f \end{bmatrix} + \mu\begin{bmatrix} a' & b' & d' \\ b' & c' & e' \\ d' & e' & f' \end{bmatrix}\right| = 0.$$

Now notice that $F(\lambda, \mu)$ is a homogeneous cubic form in $\lambda, \mu$. In turn I can apply (1.7) to $F$ to deduce:

**Proposition.** Suppose $C_{(\lambda,\mu)}$ is a pencil of conics of $\mathbb{P}^2_k$, with at least one non-degenerate conic (so that $F(\lambda, \mu)$ is not identically zero). Then the pencil has at most 3 degenerate conics. If $k = \mathbb{R}$ then the pencil has at least one degenerate conic.

**Proof.** A cubic form has $\leq 3$ zeros. Also over $\mathbb{R}$, it must have at least one zero.

**(1.14) Worked example.** Let $P_1, .. P_4$ be 4 non-collinear points of $\mathbb{P}^2_{\mathbb{R}}$; then the pencil of conics $C_{(\lambda,\mu)}$ through $P_1, .. P_4$ has 3 degenerate elements, namely the line pairs $L_{12} + L_{34}$, $L_{13} + L_{24}$, $L_{14} + L_{23}$, where $L_{ij}$ is the line through $P_i, P_j$:

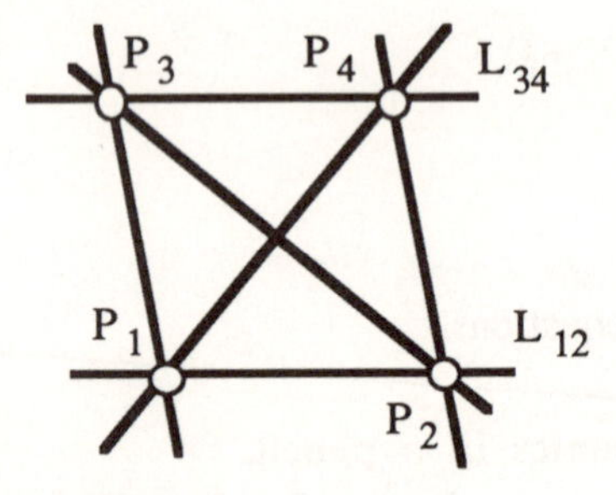


Next, suppose that I start from the pencil of conics generated by $Q_1 = Y^2 + rY + sX + t$ and $Q_2 = Y - X^2$, and try to find the points $P_1, .. P_4$ of intersection.

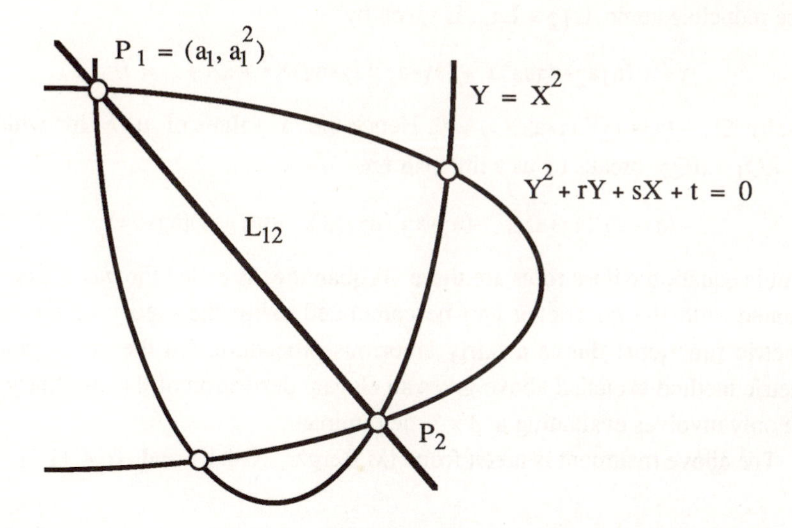


This can be done as follows: (1) find the 3 ratios $(\lambda:\mu)$ for which $C_{\lambda,\mu}$ are degenerate conics. Using what has been said above, this just means that I have to find the 3 roots of the cubic

$$F(\lambda, \mu) = \det\left|\lambda\begin{bmatrix} 0 & 0 & s/2 \\ 0 & 1 & r/2 \\ s/2 & r/2 & t \end{bmatrix} + \mu\begin{bmatrix} -1 & 0 & 0 \\ 0 & 0 & 1/2 \\ 0 & 1/2 & 0 \end{bmatrix}\right|$$

$$= -\frac{1}{4}\left(s^2\lambda^3 + (4t - r^2)\lambda^2\mu - 2r\lambda\mu^2 - \mu^3\right).$$

(2) Separate out 2 of the degenerate conics into pairs of lines (this involves solving 2 quadratic equations). (3) The 4 points $P_i$ are the points of intersection of the lines.

This procedure gives a geometric interpretation of the reduction of the general quartic in Galois theory (see for example [van der Waerden, Algebra, Ch. 8, §64]): let $k$ be a field, and $f(X) = X^4 + rX^2 + sX + t \in k[X]$ a quartic polynomial. Then the two parabolas $C_1$ and $C_2$ meet in the 4 points $P_i = (a_i, a_i^2)$ for $i = 1,.. 4$, where the $a_i$ are the 4 roots of $f$.

Then the line $L_{ij} = P_iP_j$ is given by

$$L_{ij}\colon \ (Y = (a_i + a_j)X - a_ia_j),$$

and the reducible conic $L_{12} + L_{34}$ is given by

$$Y^2 + (a_1a_2+a_3a_4)Y + (a_1+a_2)(a_3+a_4)X^2 + sX + t = 0,$$

that is, by $Q_1 - (a_1+a_2)(a_3+a_4)Q_2 = 0$. Hence the 3 values of $\mu/\lambda$ for which the conic $\lambda Q_1 + \mu Q_2$ breaks up as a line pair are

$$-(a_1+a_2)(a_3+a_4), \quad -(a_1+a_3)(a_2+a_4), \quad -(a_1+a_4)(a_2+a_3).$$

The cubic equation whose roots are these 3 quantities is called the *auxilliary cubic* associated with the quartic; it can be calculated using the theory of elementary symmetric functions; this is a fairly laborious procedure. On the other hand, the geometric method sketched above gives an elegant derivation of the auxilliary cubic which only involves evaluating a $3 \times 3$ determinant.

The above treatment is taken from [M.Berger, 16.4.10 and 16.4.11.1].

## Exercises to §1.

**1.1.** Parametrise the conic $C: (x^2 + y^2 = 5)$ by considering a variable line through $(2, 1)$ and hence find all rational solutions of $x^2 + y^2 = 5$.

**1.2.** Let $p$ be a prime; by experimenting with various $p$, guess a necessary and sufficient condition for $x^2 + y^2 = p$ to have rational solutions; prove your guess (a hint is given after Ex. 1.9 below – bet you can't do it for yourself!).

**1.3.** Prove the statement in (1.3), that an affine transformation can be used to put any conic of $\mathbb{R}^2$ into one of the standard forms (a–l). (Hint: use a linear transformation $\mathbf{x} \mapsto A\mathbf{x}$ to take the leading term $ax^2 + bxy + cy^2$ into one of $\pm x^2 \pm y^2$ or $\pm x^2$ or 0; then complete the square in $x$ and $y$ to get rid of as much of the linear part as possible.)

**1.4.** Make a detailed comparison of the affine conics in (1.3) with the projective conics in (1.6).

**1.5.** Let $k$ be any field of characteristic $\neq 2$, and $V$ a 3-dimensional $k$-vector space; let $Q: V \to k$ be a non-degenerate quadratic form on $V$. Show that if $0 \neq e_1 \in V$ satisfies $Q(e_1) = 0$ then $V$ has a basis $e_1, e_2, e_3$ such that $Q(x_1e_1 + x_2e_2 + x_3e_3) = x_1x_3 + ax_2^2$. (Hint: work with the symmetric bilinear form $\varphi$ associated to $Q$; since $\varphi$ is non-degenerate, there is a vector $e_3$ such that $\varphi(e_1, e_3) = 1$. Now find a suitable $e_2$).

Deduce that a non-empty, non-degenerate conic $C \subset \mathbb{P}^2_k$ is projectively equivalent to $(XZ = Y^2)$.

**1.6.** Let $k$ be a field with at least 4 elements, and $C: (XZ = Y^2) \subset \mathbb{P}^2_k$; prove that if $Q(X, Y, Z)$ is a quadratic form which vanishes on $C$ then $Q = \lambda(XZ - Y^2)$. (Hint: if you really can't do this for yourself, compare with the argument in (2.5).)

**1.7.** In $\mathbb{R}^3$, consider the two planes A: $(Z = 1)$ and B: $(X = 1)$; a line through 0 meeting A in $(x, y, 1)$ meets B in $(1, y/x, 1/x)$. Consider the map $\varphi: A \dashrightarrow B$ defined by $(x, y) \mapsto (y' = y/x, z' = 1/x)$; what is the image under $\varphi$ of

(i) the line $ax = y + b$; the pencil of parallel lines $ax = y + b$ (fixed a and variable b);

(ii) circles $(x-1)^2 + y^2 = c$ for variable c (distinguish the 3 cases $c > 1$, $c = 1$ and $c < 1$).

Try to imagine the above as a perspective drawing by an artist sitting at $0 \in \mathbb{R}^3$, on a plane $(X = 1)$, of figures from the plane $(Z = 1)$. Explain what happens to the points of the two planes where $\varphi$ and $\varphi^{-1}$ are undefined.

**1.8.** Let $P_1, \dots P_4$ be distinct points of $\mathbb{P}^2$ with no 3 collinear. Prove that there is a unique coordinate system in which the 4 points are $(1, 0, 0)$, $(0, 1, 0)$, $(0, 0, 1)$ and $(1, 1, 1)$. Find all conics passing through $P_1, \dots P_5$, where $P_5 = (a, b, c)$ is some other point, and use this to give another proof of (1.10) and (1.11).

**1.9.** In (1.12) there is a list of possible ways in which two conics can intersect. Write down equations showing that each possibility really occurs. (Hint: you will save yourself a lot of trouble by using symmetry and a well-chosen coordinate system.) Now find all the singular conics in the corresponding pencils.

**Hint for 1.2:** it is known from elementary number theory that $-1$ is a quadratic residue modulo p if and only if $p = 2$ or $p \equiv 1 \bmod 4$.

**1.10.** (Sylvester's determinant). Let k be an algebraically closed field, and suppose given a quadratic and cubic form in U, V as in (1.8):

$$q(U, V) = a_0U^2 + a_1UV + a_2V^2,$$

$$c(U, V) = b_0U^3 + b_1U^2V + b_2UV^2 + b_3V^3.$$

Prove that q and c have a common zero $(\eta : \tau) \in \mathbb{P}^1$ if and only if

$$\det \begin{vmatrix} a_0 & a_1 & a_2 & & \\ & a_0 & a_1 & a_2 & \\ & & a_0 & a_1 & a_2 \\ b_0 & b_1 & b_2 & b_3 & \\ & b_0 & b_1 & b_2 & b_3 \end{vmatrix} = 0.$$

(Hint: Show that if q and c have a common root then the 5 elements

$$U^2q,\ UVq,\ V^2q,\ Uc \text{ and } Vc$$

do not span the 5-dimensional vector space of forms of degree 4, and are therefore linearly dependent. Conversely, use unique factorisation in the polynomial ring k[U, V] to say something about relations of the form $Aq = Bc$ with A and B forms in (U, V), deg A =2, deg B = 1.)

**1.11.** Generalise the result of (1.10) to two forms in U, V of any degrees n and m.

## §2. Cubics and the group law

**(2.1) Examples of parametrised cubics.** Some plane cubic curves can be parametrised, just as the conics:

**Nodal cubic.** $C : (y^2 = x^3 + x^2) \subset \mathbb{R}^2$ is the image of the map $\varphi\colon \mathbb{R}^1 \to \mathbb{R}^2$ given by $t \mapsto (t^2 - 1, t^3 - t)$ (check it and see);

**Cuspidal cubic.** $C : (y^2 = x^3) \subset \mathbb{R}^2$ is the image of $\varphi\colon \mathbb{R}^1 \to \mathbb{R}^2$ given by $t \mapsto (t^2, t^3)$:

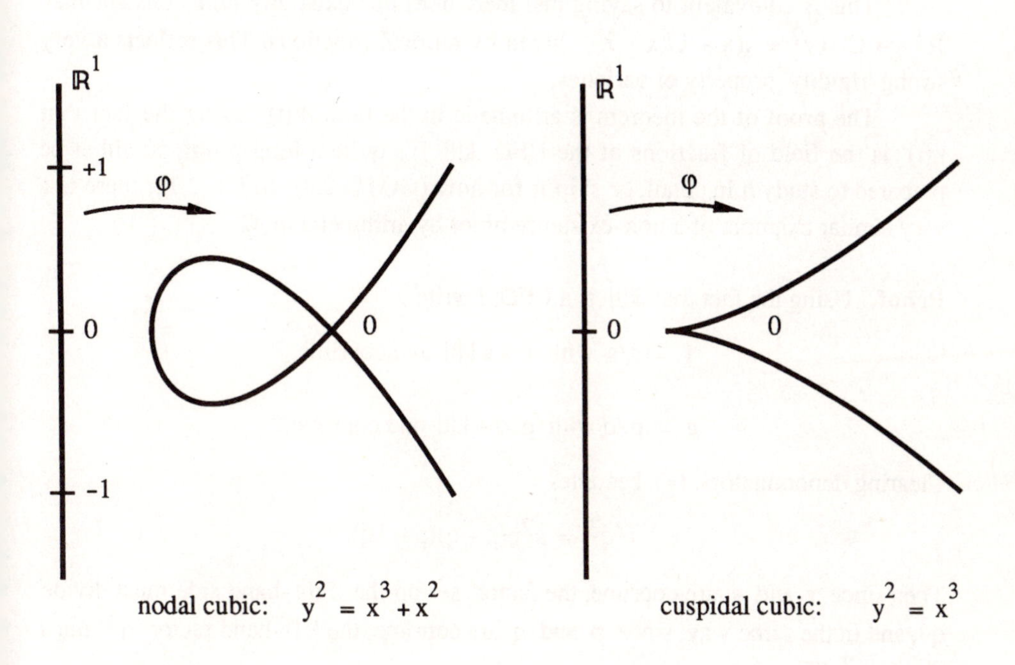

Think about the singularities of the image curve, and of the map $\varphi$. These examples will occur throughout the course, so spend some time playing with the equations; see Ex. 2.1–2.

**(2.2) The curve $(y^2 = x(x-1)(x-\lambda))$ has no rational parametrisation.** Parametrised curves are nice; for example, if you're interested in Diophantine problems, you could hope for a rule giving all $\mathbb{Q}$-valued points, as in (1.1). The parametrisation of (1.1) was of the form $x = f(t)$, $y = g(t)$, where $f$ and $g$ were *rational functions*, that is, quotients of two polynomials.

**Theorem.** Let $k$ be a field of characteristic $\neq 2$, and let $\lambda \in k$ with $\lambda \neq 0, 1$; let $f, g \in k(t)$ be rational functions such that

$$f^2 = g(g-1)(g-\lambda) \qquad (*)$$

Then $f, g \in k$.

This is equivalent to saying that there does not exist any non-constant map $\mathbb{R}^1 \dashrightarrow C : (y^2 = x(x-1)(x-\lambda))$ given by rational functions. This reflects a very strong 'rigidity' property of varieties.

The proof of the theorem is arithmetic in the field $k(t)$ using the fact that $k(t)$ is the field of fractions of the UFD $k[t]$. It's quite a long proof, so either be prepared to study it in detail, or skip it for now (GOTO 2.4). In Ex. 2.12, there is a very similar example of a non-existence proof by arithmetic in $\mathbb{Q}$.

**Proof.** Using the fact that $k[t]$ is a UFD, I write

$$f = r/s \text{ with } r, s \in k[t] \text{ and coprime,}$$

$$g = p/q \text{ with } p, q \in k[t] \text{ and coprime.}$$

Clearing denominators, $(*)$ becomes

$$r^2q^3 = s^2p(p-q)(p-\lambda q).$$

Then since $r$ and $s$ are coprime, the factor $s^2$ on the right-hand side must divide $q^3$, and in the same way, since $p$ and $q$ are coprime, the left-hand factor $q^3$ must divide $s^2$. Therefore,

$$s^2 \mid q^3 \text{ and } q^3 \mid s^2, \quad \text{so that} \quad s^2 = aq^3 \text{ with } a \in k$$

($a$ is a unit of $k[t]$, therefore in $k$).

Then

$$aq = (s/q)^2 \text{ is a square in } k[t].$$

Also,

$$r^2 = ap(p-q)(p-\lambda q),$$

so that by considering factorisation into primes, there exist non-zero constants $b, c, d \in k$ such that

$$bp, \ c(p-q), \ d(p-\lambda q)$$

are all squares in $k[t]$. If I can prove that $p, q$ are constants, then it follows from what's already been said that $r, s$ are also, proving the theorem. To prove that $p, q$ are constants, set $K$ for the algebraic closure of $k$ ; then $p, q \in K[t]$ satisfy the conditions of the next lemma.

**(2.3) Lemma.** Let $K$ be an algebraically closed field, $p, q \in K[t]$ coprime elements, and assume that 4 distinct linear combinations (that is, $\lambda p + \mu q$ for 4 distinct ratios $(\lambda : \mu) \in \mathbb{P}^1{}_K$) are squares in $K[t]$; then $p, q \in K$.

**Proof** (Fermat's method of 'infinite descent'). Both the hypotheses and conclusion of the lemma are not affected by replacing $p, q$ by

$$p' = ap + bq, \ q' = cp + dq,$$

with $a, b, c, d \in K$ and $ad - bc \neq 0$. Hence I can assume that the 4 given squares are

$$p, \ p-q, \ p-\lambda q, \ q.$$

Then $p = u^2, q = v^2$, and $u, v \in K[t]$ are coprime, with

$$\max\{\deg u, \deg v\} < \max\{\deg p, \deg q\}.$$

Now by contradiction, suppose that $\max\{\deg p, \deg q\} > 0$, and is minimal among all $p, q$ satisfying the condition of the lemma. Then both of

$$p - q = u^2 - v^2 = (u-v)(u+v)$$

and

$$p - \lambda q = u^2 - \lambda v^2 = (u - \mu v)(u + \mu v)$$

(where $\mu = \sqrt{\lambda}$) are squares in $K[t]$, so that by coprimeness of $u,v$, I conclude that each of $u - v, u + v, u - \mu v, u + \mu v$ are squares. This contradicts the minimality of $\max\{\deg p, \deg q\}$. Q.E.D.

**(2.4) Linear systems.** Write $S_d$ = {forms of degree $d$ in $(X, Y, Z)$}; (recall that a form is just a homogeneous polynomial). Any element $F \in S_d$ can be written in a unique way as

$$F = \sum a_{ijk} X^i Y^j Z^k, \quad a_{ijk} \in k,$$

with the sum taken over all $i, j, k$ with $i + j + k = d$; this means of course that $S_d$ is a $k$-vector space with basis

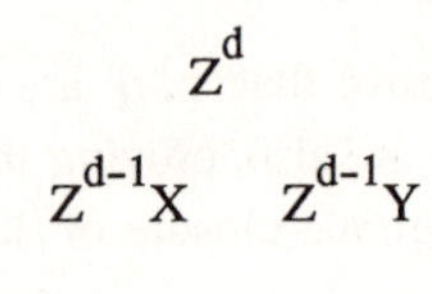

$$X^{d-1}Z \quad X^{d-2}YZ \quad .. \quad Y^{d-1}Z$$

$$X^d \quad X^{d-1}Y \quad X^{d-2}Y^2 \quad .. \quad Y^d$$

and in particular, $\dim S_d = \binom{d+2}{2}$. For $P_1, .. P_n \in \mathbb{P}^2$, let

$$S_d(P_1, .. P_n) = \{F \in S_d \mid F(P_i) = 0 \text{ for } i = 1, .. n\} \subset S_d.$$

Each of the conditions $F(P_i) = 0$, (or more precisely, $F(X_i, Y_i, Z_i) = 0$, where $P_i = (X_i : Y_i : Z_i)$) is one linear condition on $F$, so that $S_d(P_1, .. P_n)$ is a vector space of dimension $\geq \binom{d+2}{2} - n$.

**(2.5) Lemma.** Suppose that $k$ is an infinite field, and let $F \in S_d$.

(i) Let $L \subset \mathbb{P}^2_k$ be a line; if $F \equiv 0$ on $L$, then $F$ is divisible in $k[X, Y, Z]$ by the equation of $L$. That is, $F = H \cdot F'$ where $H$ is the equation of $L$ and $F' \in S_{d-1}$.

(ii) Let $C \subset \mathbb{P}^2_k$ be a non-empty non-degenerate conic; if $F \equiv 0$ on $C$, then $F$ is divisible in $k[X, Y, Z]$ by the equation of $C$. That is, $F = Q \cdot F'$ where $Q$ is the equation of $C$ and $F' \in S_{d-2}$.

If you think this statement is obvious, congratulations on your intuition: you have just guessed a particular case of the Nullstellensatz. Now find your own proof (GOTO (2.6)).

**Proof.** (i) By a change of coordinates, I can assume $L = X$. Then for any $F \in S_d$, there exists a unique expression $F = X \cdot F'_{d-1} + G(Y, Z)$: just gather together all the monomials involving $X$ into the first summand, and what's left must be a polynomial in $Y, Z$ only. Now $F \equiv 0$ on $L \Longleftrightarrow G \equiv 0$ on $L \Longleftrightarrow G(Y, Z) = 0$. The last step holds because of (1.7): if $G(Y, Z) \neq 0$ then it has at most $d$ zeros on $\mathbb{P}^1_k$, whereas if $k$ is infinite, then so is $\mathbb{P}^1_k$.

(ii) By a change of coordinates, $Q = XZ - Y^2$. Now let me prove that for any $F \in S_d$, there exists a unique expression $F = Q \cdot F'_{d-2} + A(X, Z) + YB(X, Z)$: if I just substitute $(XZ - Q)$ for $Y^2$ wherever it occurs in $F$, what's left has degree $\leq 1$ in $Y$, and is therefore of the form $A(X, Z) + YB(X, Z)$. Now as in (1.8), $C$ is the parametrised conic given by $X = U^2, Y = UV, Z = V^2$, so that

$$F \equiv 0 \text{ on } C \iff A(U^2, V^2) + UVB(U^2, V^2) \equiv 0 \text{ on } C$$

$$\iff A(U^2, V^2) + UVB(U^2, V^2) = 0 \in k[U, V] \iff A(X, Z) = B(X, Z) = 0.$$

Here the last equality comes by considering separately the terms of even and odd degrees in the form $A(U^2, V^2) + UVB(U^2, V^2)$. Q.E.D.

Ex. 2.2 gives similar cases of 'explicit' Nullstellensatz.

**Corollary.** Let $L: (H = 0) \subset \mathbb{P}^2_k$ be a line (respectively $C: (Q = 0) \subset \mathbb{P}^2_k$ a non-degenerate conic); suppose that points $P_1, .. P_n \in \mathbb{P}^2_k$ are given, and consider $S_d(P_1, .. P_n)$ for some fixed $d$. Then

(i) If $P_1, .. P_a \in L$, $P_{a+1}, .. P_n \notin L$, and $a > d$, then

$$S_d(P_1, .. P_n) = H \cdot S_{d-1}(P_{a+1}, .. P_n).$$

(ii) If $P_1, .. P_a \in C$, $P_{a+1}, .. P_n \notin C$, and $a > 2d$, then

$$S_d(P_1, .. P_n) = Q \cdot S_{d-2}(P_{a+1}, .. P_n).$$

**Proof.** (i) If $F$ is homogeneous of degree $d$, and the curve $D: (F = 0)$ meets $L$ in points $P_1, .. P_a$ with $a > d$, then by (1.8), I must have $L \subset D$, so that by the lemma, $F = H \cdot F'$; now since $P_{a+1}, .. P_n \notin L$, obviously $F' \in S_{d-1}(P_{a+1}, .. P_n)$. (ii) is exactly the same. Q.E.D.

**(2.6) Proposition.** Let $k$ be an infinite field, and $P_1,.. P_8 \in \mathbb{P}^2{}_k$ distinct points; suppose that no 4 of $P_1,.. P_8$ are collinear, and no 7 of them lie on a non-degenerate conic; then

$$\dim S_3(P_1,.. P_8) = 2.$$

**Proof.** For brevity, let me say that a set of points are *conconic* if they all lie on a non-degenerate conic. The proof of (2.6) breaks up into several cases.
**Main case.** No 3 points are collinear, no 6 conconic. This is the 'general position' case.

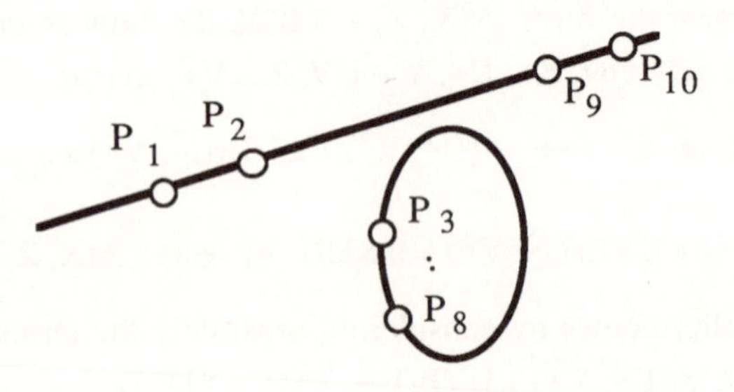


Suppose for a contradiction that $\dim S_3(P_1,.. P_8) \geq 3$, and let $P_9, P_{10}$ be distinct points on the line $L = P_1P_2$. Then

$$\dim S_3(P_1,.. P_{10}) \geq \dim S_3(P_1,.. P_8) - 2 \geq 1,$$

so that there exists $0 \neq F \in S_3(P_1,.. P_{10})$. By Corollary 2.5, $F = H{\cdot}Q$, with $Q \in S_2(P_3,.. P_8)$. Now I have a contradiction to the case assumption: if $Q$ is non-degenerate then the 6 points $P_3,.. P_8$ are conconic, whereas if $Q$ is a line-pair or a double line, then at least 3 of them are collinear.
**First degenerate case.** Suppose $P_1, P_2, P_3 \in L$ are collinear, and let $L{:}\ (H = 0)$. Let $P_9$ be a 4th point on the line $L$. Then by Corollary 2.5,

$$S_3(P_1,.. P_9) = H{\cdot}S_2(P_4,.. P_8).$$

Also, since no 4 of $P_4,.. P_8$ are collinear, by (1.9), $\dim S_2(P_4,.. P_8) = 1$, and then $\dim S_3(P_1,.. P_9) = 1$, which implies $\dim S_3(P_1,.. P_8) \leq 2$.
**Second degenerate case.** Suppose $P_1,.. P_6 \in C$ are conconic, with $C{:}\ (Q = 0)$ a non-degenerate conic. Then choose $P_9 \in C$ distinct from $P_1,.. P_6$. By Corollary 2.5 again,

$$S_3(P_1,.. P_9) = Q{\cdot}S_1(P_7, P_8);$$

the line $L = P_7P_8$ is unique, so that $S_3(P_1,.. P_9)$ is the 1-dimensional space spanned by QL, and hence $\dim S_3(P_1,.. P_8) \leq 2$. Q.E.D.

**(2.7) Corollary.** Let $C_1, C_2$ be two cubic curves whose intersection consists of 9 distinct points, $C_1 \cap C_2 = \{P_1,.. P_9\}$. Then a cubic D through $P_1,.. P_8$ also passes through $P_9$.

**Proof.** If 4 of the points $P_1,.. P_9$ were on a line L, then each of $C_1$ and $C_2$ would meet L in $\geq 4$ points, and thus contain L, which contradicts the assumption on $C_1 \cap C_2$. For exactly the same reason, no 7 of the points can be conconic. Therefore the assumptions of (2.6) are satisfied, so I can conclude that

$$\dim S_3(P_1,.. P_8) = 2;$$

this means that the equations $F_1, F_2$ of $C_1, C_2$ form a basis of $S_3(P_1,.. P_8)$, and hence D: $(G = 0)$, where $G = \lambda F_1 + \mu F_2$. Now $F_1, F_2$ vanish at $P_9$, hence so does G. Q.E.D.

**(2.8) Group law on a plane cubic.** Suppose $k \subset \mathbb{C}$ is a subfield of $\mathbb{C}$, and $F \in k[X, Y, Z]$ a cubic form defining a (non-empty) plane curve $C: (F = 0) \subset \mathbb{P}^2_k$. Assume that F satisfies the following two conditions:

(a) F is irreducible (so that C does not contain a line or conic);

(b) for every point $P \in C$, there exists a unique line $L \subset \mathbb{P}^2_k$ such that P is a repeated zero of $F_{|L}$.

Note that geometrically, the condition in (b) is that C should be non-singular, and the line L referred to is the tangent line $L = T_PC$ (see Ex. 2.3). This will be motivation for the general definition of non-singularity and tangent spaces to a variety in §6.

Fix any point $O \in C$, and make the following construction:

**Construction.** (i) For $A \in C$, let $\overline{A}$ = 3rd point of intersection of C with the line OA;

(ii) for $A, B \in C$, write R = 3rd point of intersection of AB with C, and define $A + B$ by $A + B = \overline{R}$ (see diagram below).

**Theorem.** The above construction defines an Abelian group law on C, with O as zero (= neutral element).

**Proof.** Associativity is the crunch here; I start the proof by first clearing up the easy bits.

(I) I have to prove that the addition and inverse operations are well-defined; if $P, Q \in C$ are any two points, then either $P \neq Q$, and the line $L = PQ \subset \mathbb{P}^2_k$ is uniquely determined, or $P = Q$ then by the assumption (b), there is a unique line $L \subset \mathbb{P}^2_k$ such that $P$ is a repeated zero of $F_{|L}$; in either case, $F_{|L}$ is a cubic form in two variable, having 2 given k-valued zeros. It therefore splits as a product of 3 linear factors, and hence without exception, the 3rd residual point of intersection $R$ is defined and has coordinates in k. Note that any of $P = Q$, $P = R$, $Q = R$, or $P = Q = R$ are allowed; these correspond algebraically to $F_{|L}$ having multiple zeros, and geometrically to tangent and inflexion points.

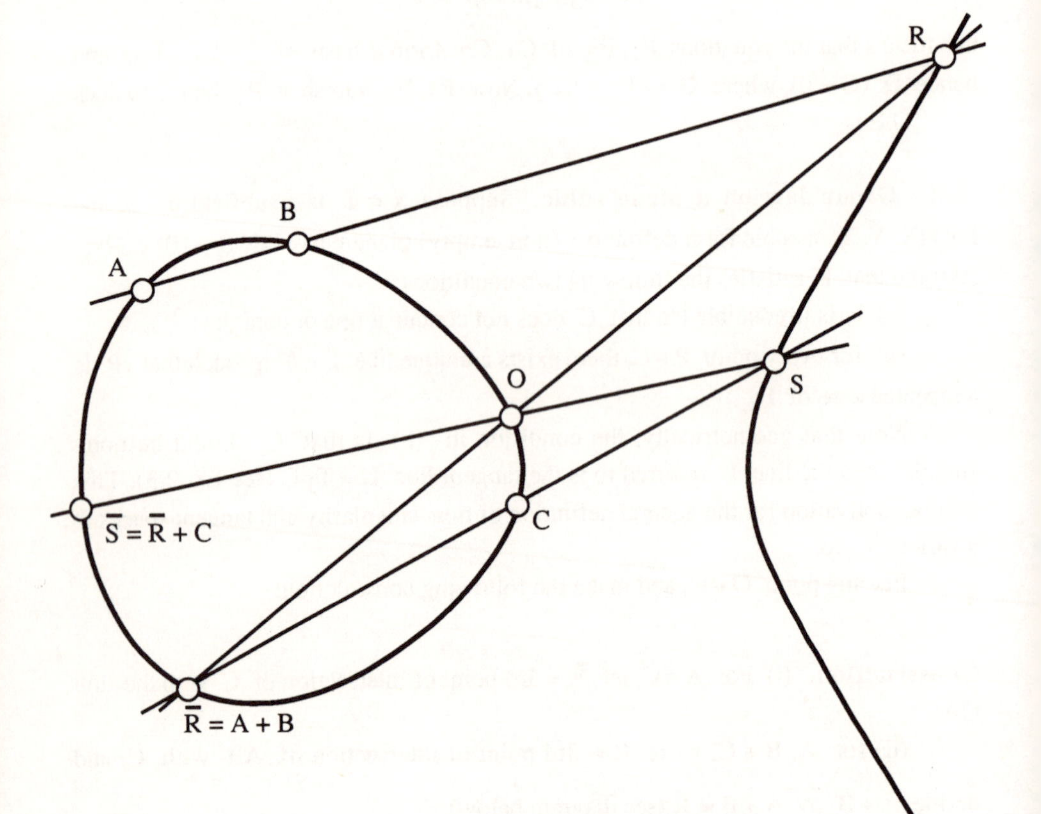


Cubic curve and its group law

(II) Verifying that the given point $O$ is the neutral element is completely formal: since $OA\bar{A}$ are collinear, the construction of $O + A$ consists of taking the line $L = OA$ to get the 3rd point of intersection $\bar{A}$, then the same line $L = O\bar{A}$ to get back to $A$.

(III) I think I'll leave $A + B = B + A$ to the reader.

(IV) To find the inverse, first define the point $\bar{O}$ as in (i) of the construction: let $L$ be the line such that $F_{|L}$ has $O$ as a repeated zero, and define $\bar{O}$ to be the 3rd point of intersection of $L$ with $C$; then it is easy to check that the 3rd point of intersection of $\bar{O}A$ with $C$ is the inverse of $A$ for every $A \in C$.

**(2.9)** Now I give the proof of associativity for 'sufficiently general' points: suppose that $A, B, C$ are 3 given points of $C$; then the construction of $(A + B) + C = \bar{S}$ uses 4 lines (see diagram above)

$$L_1 : ABR,\ L_2 : RO\bar{R},\ L_3 : C\bar{R}S,\ L_4 : SO\bar{S}.$$

The construction of $(B + C) + A = \bar{S}'$ uses 4 lines

$$M_1 : BCQ,\ M_2 : QO\bar{Q},\ M_3 : A\bar{Q}S' \text{ and } M_4 : S'O\bar{S}'.$$

I want to prove $\bar{S} = \bar{S}'$, and clearly for this it is enough to prove $S = S'$; to do this, consider the 2 cubics

$$D_1 = L_1 + M_2 + L_3 \text{ and } D_2 = M_1 + L_2 + M_3.$$

Then by construction,

$$C \cap D_1 = \{A, B, C, O, R, \bar{R}, Q, \bar{Q}, S\},$$

and

$$C \cap D_2 = \{A, B, C, O, R, \bar{R}, Q, \bar{Q}, S'\}.$$

Now provided the 9 points $\{A, B, C, O, R, \bar{R}, Q, \bar{Q}, S\}$ are all distinct, the two cubics $C$ and $D_1$ satisfy the conditions of Proposition 2.7; therefore, $D_2$ must pass through $S$, and the only way that this can happen is if $S = S'$.

There are several ways to complete the argument. The most thorough of these gives a genuine treatment of the intersection of two curves taking into account multiple intersections (roughly, in terms of 'ideals of intersection'), and the statement corresponding to (2.7) is Max Noether's Lemma (see [Fulton, p.120 and p.124]).

**(2.10)** I sketch one version of the argument 'by continuity', which uses the fact that $k \subset \mathbb{C}$. Write $C_{\mathbb{C}} \subset \mathbb{P}^2_{\mathbb{C}}$ for the complexified curve $C$, that is, the set of ratios $(X : Y : Z)$ of complex numbers satisfying the same equation $F(X, Y, Z) = 0$. If the associative law holds for all $A, B, C \in C_{\mathbb{C}}$, then obviously also for all points in $C$. Therefore, I can assume that $k = \mathbb{C}$.

The reader who cares about it will have no difficulty in finding proofs of the following two statements (see Ex. 2.8):

**Lemma.** (i) $A + B$ is a continuous function of $A$ and $B$;

(ii) for all $A, B, C \in C$ there exist $A', B', C' \in C$ arbitrarily near to $A, B, C$ such that the 9 points $\{A', B', C', O, R, \overline{R}, Q, \overline{Q}, S\}$ constructed from them are all distinct.

The addition law is a map $\varphi\colon C \times C \to C$ given by $(A, B) \mapsto A + B$. By (i), $\varphi$ is continuous, and hence so are the two maps (sorry!)

$$f = \varphi \circ (\varphi \times \mathrm{id}_C) \quad \text{and} \quad g = \varphi \circ (\mathrm{id}_C \times \varphi)\colon C \times C \times C \to C$$

given by $(A, B, C) \mapsto (A + B) + C$ and $A + (B + C)$. Also, by (ii), the subset $U \subset C \times C \times C$ consisting of triples $(A, B, C)$ for which the 9 points of the construction are distinct is dense; by the above argument, $f$ and $g$ thus coincide on $U$, and since they are continuous, they coincide everywhere. Q.E.D.

**Remark.** The continuity argument as it stands involves the topology of $\mathbb{C}$, and is thus not purely algebraic. In fact the addition map $\varphi$ is a morphism of varieties $\varphi\colon C \times C \to C$, as will be proved later (see (4.14)), and the remainder of the argument can also be reformulated in this purely algebraic form: the subset of $C \times C \times C$ for which the 9 points are distinct is a dense open set for the Zariski topology, and two morphisms which coincide on a dense open set coincide everywhere. (I hope that this remark can provide useful motivation for the rest of the course; if you find it confusing, just ignore it for the moment.)

**(2.11) Pascal's Theorem** (the mystic hexagon). The diagram

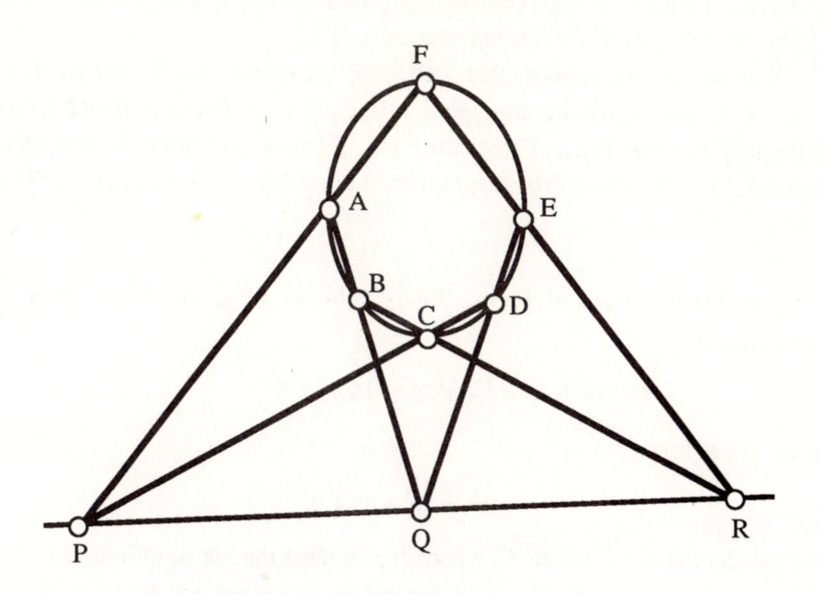


consists of a hexagon ABCDEF in $\mathbb{P}^2{}_k$ with pairs of opposite sides extended until they meet in points P, Q, R. Assume that the nine points and the six lines of the diagram are all distinct; then

$$\text{ABCDEF are conconic} \iff \text{PQR are collinear.}$$

This famous theorem is a rather similar application of (2.7), and is given just for fun; of course, other proofs are possible, see any text on geometry, for example [Berger, 16.2.10 and 16.8.3–5].

**Proof.** In the diagram consider the two triples of lines

$$L_1\colon \text{PAF},\ L_2\colon \text{QDE},\ L_3\colon \text{RBC},$$

and

$$M_1\colon \text{PCD},\ M_2\colon \text{QAB},\ M_3\colon \text{REP};$$

let $C_1 = L_1 + L_2 + L_3$ and $C_2 = M_1 + M_2 + M_3$. Now I'm all set to apply (2.7), since clearly $C_1$ and $C_2$ are two cubics such that

$$C_1 \cap C_2 = \{A, B, C, D, E, F, P, Q, R\}.$$

Suppose PQR are collinear, with $L = PQR$; let $\Gamma$ be the conic through ABCDE (the existence and unicity of which is provided by (1.9)). Then by construction, $L + \Gamma$ is a cubic passing through the 8 points {A, B, C, D, E,

P, Q, R}, and by (2.7), it must contain F; by assumption, $F \notin L$, so that necessarily $F \in \Gamma$, proving that the six points are conconic.

Now conversely, suppose that ABCDEF are on a conic $\Gamma$, and let $L = PQ$; then $L + \Gamma$ is a cubic passing through {A, B, C, D, E, F, P, Q}, so by (2.7) it must pass through R. Now R can't be on the conic $\Gamma$ (since otherwise $\Gamma$ is a line pair, and some of the 6 lines of the diagram must concide), so $R \in L$, that is, PQR are collinear. Q.E.D.

**(2.12) Inflexion, normal form.** Every cubic in $\mathbb{P}^2_{\mathbb{R}}$ or $\mathbb{P}^2_{\mathbb{C}}$ can be put in the normal form

$$C: \; Y^2Z = X^3 + aXZ^2 + bZ^3, \qquad (**)$$

or in the affine form

$$y^2 = x^3 + ax + b.$$

Now consider the above curve C; where does it meet the line at infinity L: (Z = 0)? That's easy, just substitute $Z = 0$ in the defining polynomial $F = -Y^2Z + X^3 + aXZ^2 + bZ^3$ to get $F_{|L} = X^3$; this means that $F_{|L}$ has a triple zero at $P = (0, 1, 0)$. To see what this means geometrically, set $Y = 1$, to get the equation in affine coordinates (x, z) around (0, 1, 0):

$$z = x^3 + axz^2 + bz^3.$$

This curve is approximated to a high degree of accuracy by $z = x^3$:

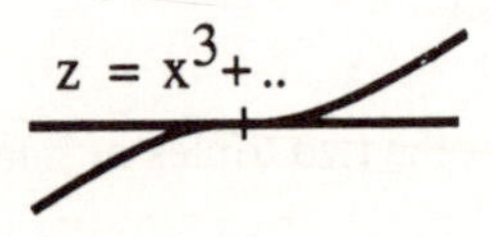


the behaviour is described by saying that C has an *inflexion point* at (0, 1, 0). More generally, an inflexion point P on a curve C is defined by the condition that there is a line $L \subset \mathbb{P}^2_k$ such that $F_{|L}$ has a zero of multiplicity $\geq 3$ at P (see Ex. 2.9; in fact necessarily $L = T_PC$ by (2.8, (b)), and the multiplicity $= 3$ by (1.9)). It is not hard to interpret this in terms of the derivatives and second derivatives of the defining equations: for example, if the defining equation is $y = f(x)$, then the condition for an inflexion point is simply $d^2f/dx^2(P) = 0$; this corresponds in the diagram to the curve passing through a transition from being 'concave downwards' to being 'concave upwards'.

It can be shown (see Ex. 2.10) that conversely, if a plane cubic C has an inflexion point, then its equation can be put in normal form (∗∗) as above.

**(2.13) Simplified group law.** The normal form (∗∗) is extremely convenient for the group law: take the inflexion point $O = (0, 1, 0)$ as the neutral element. Under these conditions, the group law becomes particularly nice, for the following reasons:

(a) $C = \{O\} \cup$ affine curve $C_0$: $(y^2 = x^3 + ax + b)$;

so it is legitimate to treat $C$ as an affine curve, with occasional references to the single point $O$ at infinity, the $0$ of the group law.

(b) The lines through $O$, which are the main ingredient in part (i) of the construction of the group law in (2.7), are given projectively by $X = \lambda Z$, and affinely by $x = \lambda$; any such line meets $C$ at points $(\lambda, \pm\sqrt{(\lambda^3 + a\lambda + b)})$, and at infinity. Hence if $P = (x, y)$, then the point $\overline{P}$ constructed in (2.8, (i)) is $(x, -y)$; thus $P \mapsto \overline{P}$ is the natural symmetry $(x, y) \mapsto (x, -y)$ of the curve $C_0$:

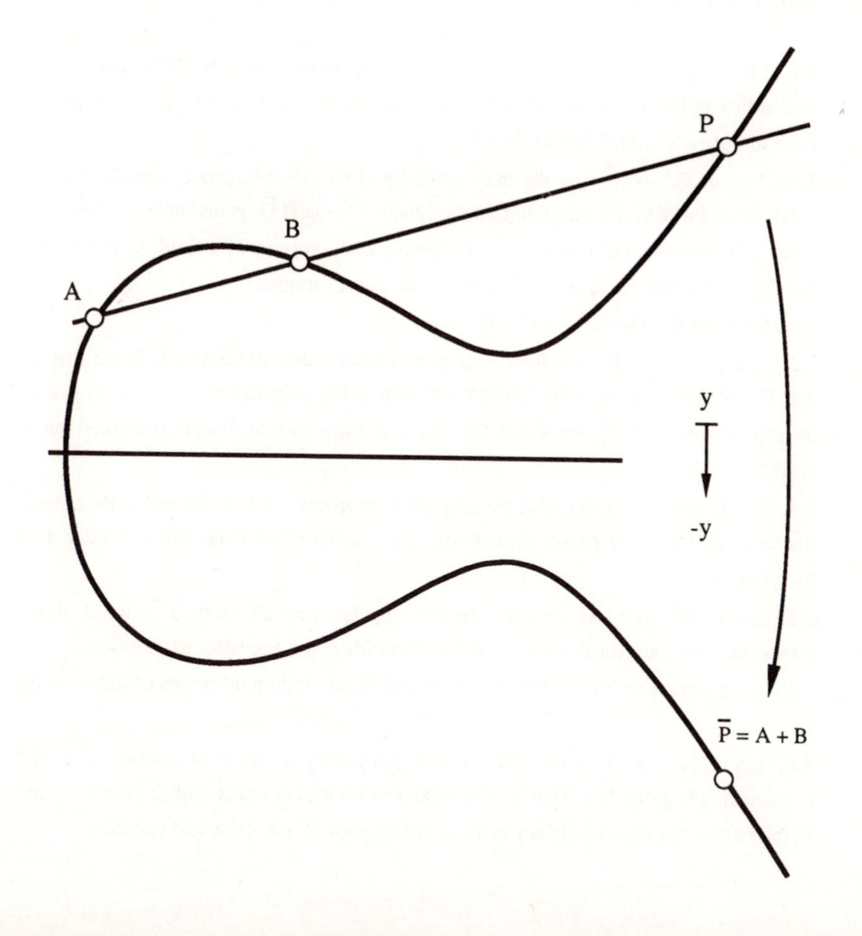

(c) The inverse of the group law (2.9, (IV)) is described in terms of the point $\overline{O}$, constructed as the 3rd point of intersection of the unique line $L$ such that $F_{|L}$ has $O$ as a repeated zero; however, in our case, this line is the line at infinity $L: (Z = 0)$, and $L \cap C = 3O$, so that $\overline{O} = O$, and the inverse of the group law then simplifies to $-P = \overline{P}$.

I can now restate the group law as a much simplified version of Theorem 2.8:

**Theorem.** Let $C$ be a cubic in the normal form (**); then there is a unique group law on $C$ such that $O = (0, 1, 0)$ is the neutral element, the inverse is given by $(x, y) \mapsto (x, -y)$, and for all $P, Q, R \in C$,

$$P + Q + R = O \iff P, Q, R \text{ are collinear.}$$

## Exercises to §2.

**2.1.** Let $C: (y^2 = x^3 + x^2) \subset \mathbb{R}^2$. Show that a variable line through $(0,0)$ meets $C$ at one further point, and hence deduce the parametrisation of $C$ given in (2.1). Do the same for $(y^2 = x^3)$ and $(x^3 = y^3 - y^4)$.

**2.2.** Let $\varphi: \mathbb{R}^1 \to \mathbb{R}^2$ be the map given by $t \mapsto (t^2, t^3)$; prove directly that any polynomial $f \in \mathbb{R}[X, Y]$ vanishing on the image $C = \varphi(\mathbb{R}^1)$ is divisible by $Y^2 - X^3$. (Hint: use the method of Lemma 2.5.) Determine what property of a field $k$ will ensure that the result holds for $\varphi: k \to k^2$ given by the same formula.

Do the same for $t \mapsto (t^2 - 1, t^3 - t)$.

**2.3.** Let $C: (f = 0) \subset k^2$, and let $P = (a, b) \in C$; assume that $\partial f/\partial x(P) \neq 0$. Prove that the line $L: (\partial f/\partial x(P)\cdot(X - a) + \partial f/\partial y(P)\cdot(Y - b) = 0)$ is the tangent line to $C$ at $P$, that is, the unique line $L$ of $k^2$ for which $f_{|L}$ has a multiple root at $P$ (this is worked out in detail in (6.1)).

**2.4.** Let $C: (y^2 = x^3 + 4x)$, with the simplified group law (2.12). Show that the tangent line to $C$ at $P = (2, 4)$ passes through $(0, 0)$, and deduce that $P$ is a point of order 4 in the group law.

**2.5.** Let $C: (y^2 = x^3 + ax + b) \subset \mathbb{R}^2$ be non-singular; find all points of order 2 in the group law, and understand what group they form (there are two cases to consider).

Now explain geometrically how you would set about finding all points of order 4 on $C$.

**2.6.** Let $C: (y^2 = x^3 + ax + b) \subset \mathbb{R}^2$; write a computer program to sketch part of $C$, and to calculate the group law. That is, it prompts you for the coordinates of 2 points $A$ and $B$, then draws the lines and tells you the coordinates of $A + B$. (Use real variables.)

**2.7.** Let $C: (y^2 = x^3 + ax + b) \subset k^2$; if $A = (x_1, y_1)$ and $B = (x_2, y_2)$, show how to give the coordinates of $A + B$ as rational functions of $x_1, y_1, x_2, y_2$. (Hint: if $F(X)$ is a polynomial of degree 3 and you know 2 of the roots, you can find the 3rd by looking at just one coefficient of $F$. This is a question with a non-unique answer, since there are many correct expressions for the rational functions. One solution is given in (4.14).)

**2.8.** By writing down the equation of the tangent line to $C$ at $A$, find a formula for $2A$ in the group law on $C$, and verify that it is the limit of a suitable formula for $A + B$ as $B$ tends to $A$. (Hint. Use Ex. 2.7 and if necessary refer to (4.14).)

**2.9.** Let $x, z$ be coordinates on $k^2$, and let $f \in k[x, z]$; write $f$ as

$$f = a + bx + cz + dx^2 + exz + fz^2 + \dots$$

Write down the conditions in terms of a, b, c,.. that must hold in order that

(i) $P = (0, 0) \in C: (f = 0)$;

(ii) the tangent line to $C$ at $P$ is $(z = 0)$;

(iii) $P$ is an inflexion point of $C$ with $(z = 0)$ as the tangent line.

(Recall from (2.12) that $P \in C$ is an inflexion point if the tangent line $L$ is defined, and $f_{|L}$ has at least a 3-fold zero at $P$.)

**2.10.** Let $C \subset \mathbb{P}^2_k$ be a plane cubic, and suppose that $P \in C$ is an inflexion point; prove that a change of coordinates in $\mathbb{P}^2_k$ can be used to bring $C$ into the normal form $(Y^2Z = X^3 + aX^2Z + bXZ^2 + cZ^3)$. (Hint: take coordinates such that $P = (0,1,0)$ and the inflexion tangent is $(Z = 0)$; then using the previous question, in local coordinates $(x, z)$, $Y$ will appear in a quadratic term $Y^2Z$, and otherwise only linearly. Show then that you can get rid of the linear term in $Y$ by completing the square.)

**2.11.** (Group law on a cuspidal cubic.) Consider the curve $C: (z = x^3) \subset k^2$; $C$ is the image of the bijective map $\varphi: k \to C$ by $t \mapsto (t, t^3)$, so it inherits a group law from the additive group $k$. Prove that this is the unique group law on $C$ such that $(0, 0)$ is the neutral element and

$$P + Q + R = 0 \iff P, Q, R \text{ are collinear}$$

for $P, Q, R \in C$. (Hint: you might find useful the identity

$$\det \begin{vmatrix} 1 & a & a^3 \\ 1 & b & b^3 \\ 1 & c & c^3 \end{vmatrix} = (a - b)(b - c)(c - a)(a + b + c).)$$

In projective terms, $C$ is the curve $(Y^2Z = X^3)$, our old friend with a cusp at the origin and an inflexion point at $(0, 1, 0)$, and the point of the question is that the usual construction gives a group law on the complement of the singular point.

**2.12.** (Due to Leonardo Pisano, known as Fibonacci, A.D.1220.) Prove that for $u, v \in \mathbb{Z}$,

$$u^2 + v^2 \text{ and } u^2 - v^2 \text{ both squares} \implies v = 0.$$

Hints (due to P. de Fermat, see J.W.S.Cassels, Journal of London Math Soc. 41 (1966), p.207):

**Step 1.** Reduce to solving

$x^2 = u^2 + v^2,\ y^2 = u^2 - v^2$ with $x, y, u, v \in \mathbb{Z}$ pairwise coprime. (*)

**Step 2.** Considerations mod 4 show that $x, y, u$ are odd and $v$ even.

**Step 3.** The 4 pairs of factors on the l.-h.s. of the factorisations

$$\begin{aligned} (x-u)(x+u) &= v^2 \\ (u-y)(u+y) &= v^2 \\ (x-y)(x+y) &= 2v^2 \\ (2u-x-y)(2u+x+y) &= (x-y)^2 \end{aligned} \qquad (**)$$

have no common factor other than 2.

**Step 4.** Replacing $y$ by $-y$ if necessary, we can assume that $4 \nmid x - y$. Now by considering the parity of factors on l.-h.s. of (**), prove that

$$x - u = 2u_1^2,\ u - y = 2v_1^2,\ x - y = 2x_1^2 \text{ and } 2u - x - y = 2y_1^2$$

with $u_1, v_1, x_1, y_1 \in \mathbb{Z}$.

**Step 5.** Show that $u_1, v_1, x_1, y_1$ is another solution of (*) with $v_1 < v$, and deduce a contradiction by 'infinite descent'.

Compare this argument with the proof of (2.2), which was easier only in that I didn't have to mess about with 2's.

# Appendix to Chapter 1
# Curves and their genus

**(2.14) Topology of a non-singular cubic.** It is easy to see that a non-singular plane cubic $C: (y^2 = x^3 + ax + b) \subset \mathbb{P}^2_{\mathbb{R}}$ has one of the two shapes

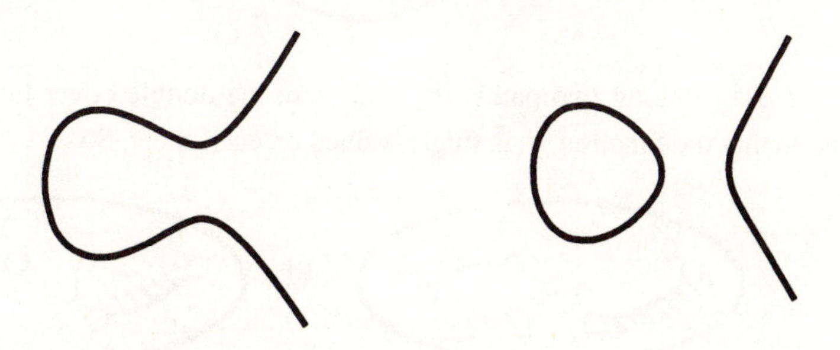

That is, topologically, $C$ is either one or two circles (including the single point at infinity, of course). To look at the same question over $\mathbb{C}$, take the alternative normal form

$$C: (y^2 = x(x-1)(x-\lambda)) \cup \{\infty\};$$

what is the topology of $C \subset \mathbb{P}^2_{\mathbb{C}}$? The answer is a torus:

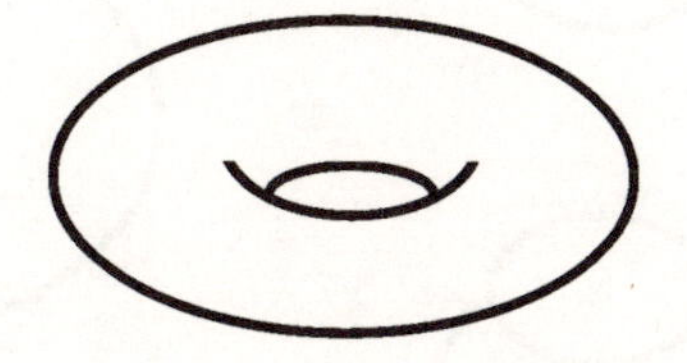

The idea of the proof is to consider the map

$$\pi: C \to \mathbb{P}^1_{\mathbb{C}} \text{ by } (X, Y, Z) \mapsto (X, Z), \text{ and } \infty \mapsto (1, 0);$$

in affine coordinates this is $(x, y) \mapsto x$, so it's the 2-to-1 map corresponding to the graph of $y = \pm\sqrt{(x(x-1)(x-\lambda))}$. Everyone knows that $\mathbb{P}^1_{\mathbb{C}}$ is homeomorphic to $S^2$, the Riemann sphere ('stereographic projection'); consider the 'function' $y(x) = \pm\sqrt{(x(x-1)(x-\lambda)}$ on $\mathbb{P}^1_{\mathbb{C}}$. This is 2-valued outside $\{0, 1, \lambda, \infty\}$:

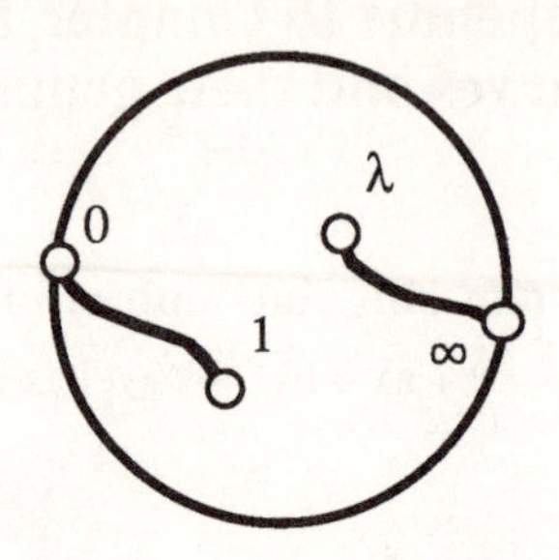


Now cut $\mathbb{P}^1_{\mathbb{C}}$ along two paths 01 and $\lambda\infty$; the double cover falls apart as 2 pieces, so that the function y is single-valued on each sheet. So

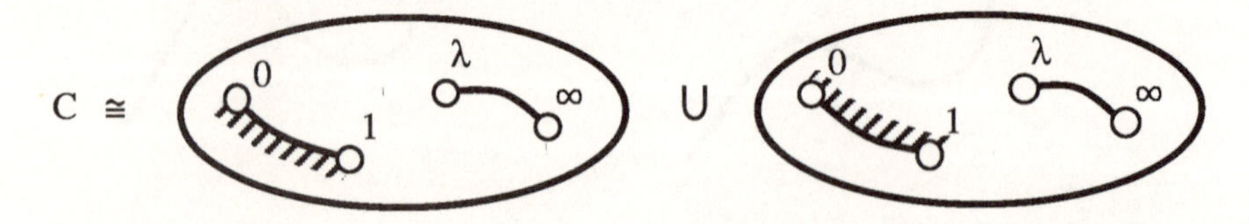


(the shading indicates how the two sheets match up under the glueing). To see what's going on, open up the slits:

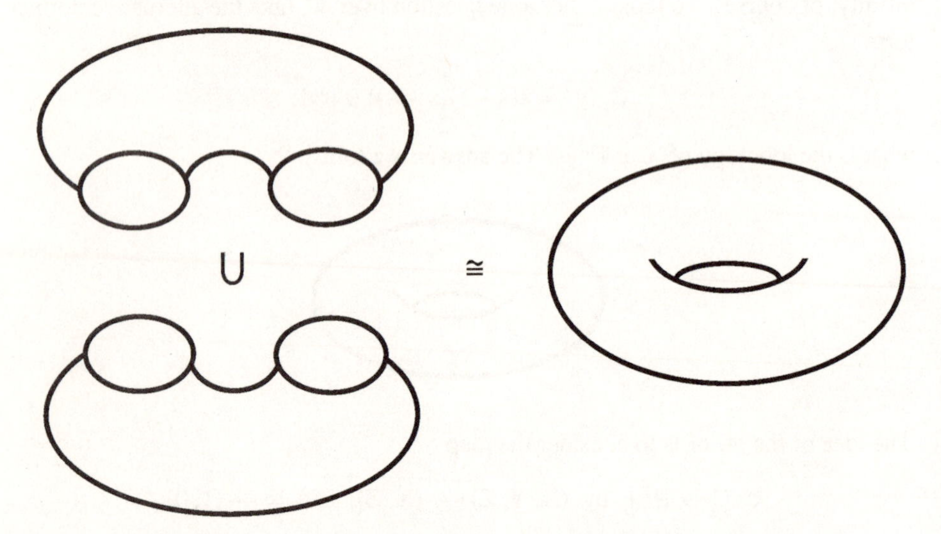


**(2.15) Discussion of genus.** A non-singular projective curve C over $\mathbb{C}$ has got just one topological invariant, its genus $g = g(C)$:

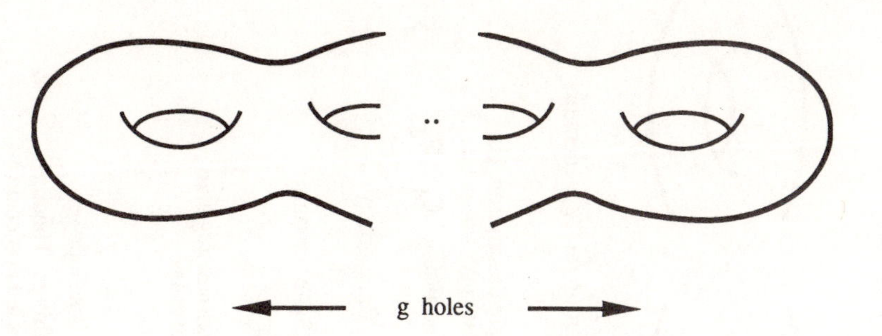


For example, the affine curve $C: (y^2 = f_{2g+1}(x) = \prod_i (x - a_i)) \subset \mathbb{C}^2$, where $f_{2g+1}$ is a polynomial of degree $2g+1$ in $x$ with distinct roots $a_i$, can be related to the Riemann surface of $\sqrt{f}$ exactly as in (2.13), and be viewed as a double cover of the Riemann sphere $\mathbb{P}^1_{\mathbb{C}}$ branched in the $2g+1$ points $a_i$ and in $\infty$, and by the same argument, can be seen to have genus $g$. As another example, the genus of a nonsingular plane curve $C_d \subset \mathbb{P}^2_{\mathbb{C}}$ of degree $d$ is given by

$$g = g(C_d) = \binom{d-1}{2}.$$

**(2.16) Commercial break.** Complex curves (= compact Riemann surfaces) appear across a whole spectrum of math problems, from Diophantine arithmetic through complex function theory and low-dimensional topology to differential equations of math physics. So go out and buy a complex curve today.

To a quite extraordinary degree, the properties of a curve are determined by its genus, and more particularly by the trichotomy $g = 0$, $g = 1$ or $g \geq 2$. Some of the more striking aspects of this are described in the table on the following page, and I give a brief discussion; this ought to be in the background culture of every mathematician.

To give a partial answer to the Diophantine question mentioned in (1.1-2) and again in (2.1), it is known that a curve can be parametrised by rational functions if and only if $g = 0$; if I'm working over a fixed field, a curve of genus 0 may have no $k$-valued points at all (for example, the conic in (1.2)), but if it has one point, it can be parametrised over $k$, so that its $k$-valued points are in bijection with $\mathbb{P}^1_k$. Any curve of genus 1 is isomorphic to a cubic as in this section, and a group law is defined on the $k$-valued points (provided of course that there exists at least one - there's no such thing as the empty group); if $k$ is a number field (for example, $k = \mathbb{Q}$), the $k$-valued points form an Abelian group which is finitely generated (the

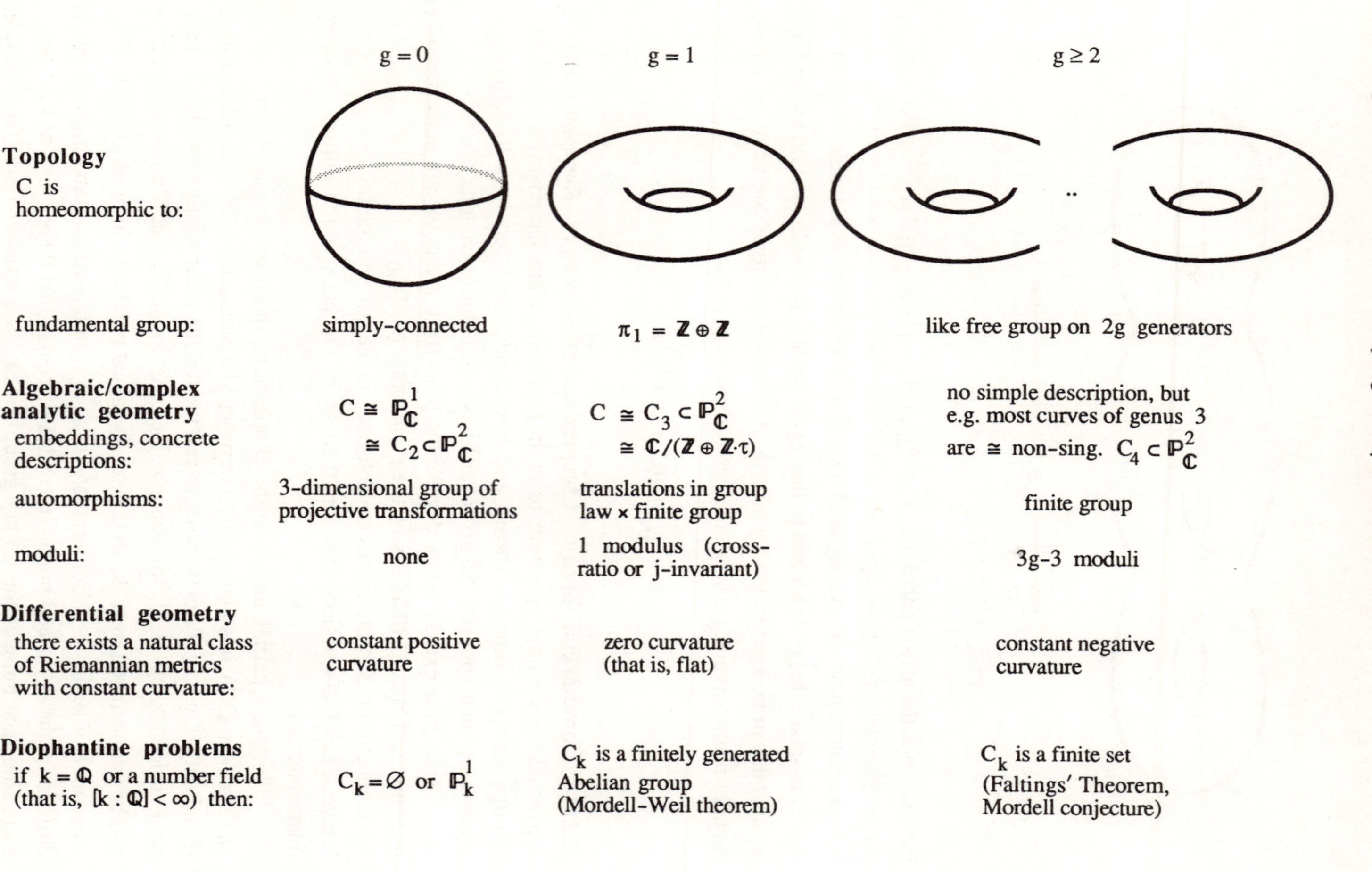

| | $g=0$ | $g=1$ | $g\geq 2$ |
|---|---|---|---|
| **Topology**<br>C is homeomorphic to: | | | |
| fundamental group: | simply-connected | $\pi_1 = \mathbb{Z}\oplus\mathbb{Z}$ | like free group on 2g generators |
| **Algebraic/complex analytic geometry**<br>embeddings, concrete descriptions: | $C \cong \mathbb{P}^1_{\mathbb{C}}$<br>$\cong C_2 \subset \mathbb{P}^2_{\mathbb{C}}$ | $C \cong C_3 \subset \mathbb{P}^2_{\mathbb{C}}$<br>$\cong \mathbb{C}/(\mathbb{Z}\oplus\mathbb{Z}\cdot\tau)$ | no simple description, but e.g. most curves of genus 3 are $\cong$ non-sing. $C_4 \subset \mathbb{P}^2_{\mathbb{C}}$ |
| automorphisms: | 3-dimensional group of projective transformations | translations in group law × finite group | finite group |
| moduli: | none | 1 modulus (cross-ratio or j-invariant) | 3g-3 moduli |
| **Differential geometry**<br>there exists a natural class of Riemannian metrics with constant curvature: | constant positive curvature | zero curvature (that is, flat) | constant negative curvature |
| **Diophantine problems**<br>if $k=\mathbb{Q}$ or a number field (that is, $[k:\mathbb{Q}]<\infty$) then: | $C_k=\emptyset$ or $\mathbb{P}^1_k$ | $C_k$ is a finitely generated Abelian group (Mordell-Weil theorem) | $C_k$ is a finite set (Faltings' Theorem, Mordell conjecture) |

Mordell-Weil Theorem). Whereas a curve of genus $g \geq 2$ is now known to have only a finite set of k-valued points; this is a famous theorem proved by Faltings in 1983, and for which he received the Fields medal in 1986. Thus for example, for any $n \geq 4$, the Fermat curve $x^n + y^n = 1$ has at most a finite number of rational points.

Over $\mathbb{C}$, a curve C of genus 1 is topologically a torus, and has a group law, so that it is analytically of the form $C \cong \mathbb{C}/(\mathbb{Z} \oplus \mathbb{Z}\cdot\tau)$:

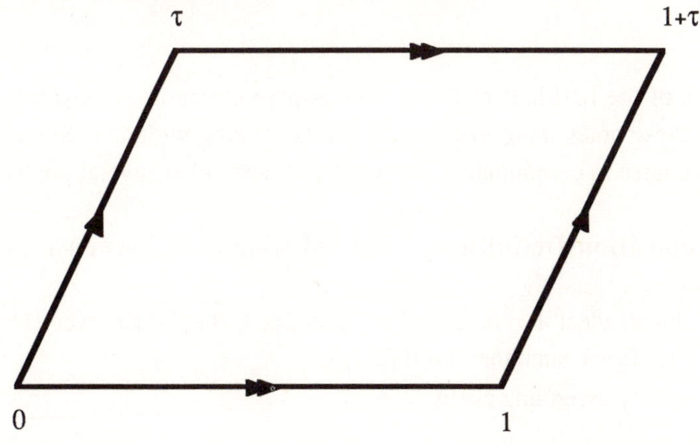


The isomorphism between this quotient and a plane curve $C_3 \subset \mathbb{P}^2_{\mathbb{C}}$ is given by a holomorphic map $\varphi\colon \mathbb{C} \to C_3$, that is, a kind of 'parametrisation' of $C_3$; but $\varphi$ cannot be in terms of rational functions (by (2.3)), and is $\infty$-to-1; this is the theory of doubly periodic functions of a complex variable, which was one of the mainstays of 19th century analysis (Weierstrass $\wp$-function, Riemann theta-function).

Another important thing to notice is that different periods $\tau$ will usually lead to different curves; they're all homeomorphic to the standard torus $S^1 \times S^1$, but as algebraic curves, or complex analytic curves, they're not isomorphic. The period $\tau$ is a *modulus*, that is, a complex parameter which governs variation of the complex structure C on the fixed topological object $S^1 \times S^1$.

The student interested in more on curves should look at [D. Mumford, Curves and their Jacobians], the first part of which is fairly colloquial, or [Clemens].

# Chapter II. The category of affine varieties

## §3. Affine varieties and the Nullstellensatz

Much of the first half of this section is pure commutative algebra; note that throughout these notes, *ring* means a commutative ring with a 1. Since this is not primarily a course in commutative algebra, I will hurry over several points.

**(3.1) Proposition-Definition.** The following conditions on a ring A are equivalent.

(i) Every ideal $I \subset A$ is finitely generated; that is, for every ideal $I \subset A$, there exist $f_1,.. f_k \in I$ such that $I = (f_1,.. f_k)$.

(ii) Every ascending chain

$$I_1 \subset .. I_m \subset ..$$

of ideals of A terminates, that is the chain is eventually stationary, with $I_N = I_{N+1} = ..$ (the *ascending chain condition*, or *a.c.c.* ).

(iii) Every non-empty set of ideals of A has a maximal element.

If they hold, A is a *Noetherian* ring.

**Proof.** (i) ⟹ (ii) Given $I_1 \subset .. I_m \subset ..$, set $I = \bigcup I_m$. Then clearly I is still an ideal. If $I = (f_1, .. f_k)$, then each $f_i$ is an element of some $I_{m(i)}$ for some m(i), so that taking $m = \max(m(i))$ gives $I = I_m$, and the chain stops at $I_m$.

(ii) ⟹ (iii) is clear. (Actually, it uses the axiom of choice.)

(iii) ⟹ (i) Let I be any ideal; write $\Sigma = \{J \subset I \mid J \text{ is f.g. ideal}\}$. Then by (iii), Σ has a maximal element, say $J_0$. But then $J_0 = I$, because otherwise any $f \in I \setminus J_0$ gives an ideal $J_0 + Af$ which is still finitely generated, but strictly bigger than $J_0$. Q.E.D.

As a thought experiment, prove that $\mathbb{Z}$ and k[X] are Noetherian.

**(3.2) Proposition.** (i) Suppose that A is Noetherian, and $I \subset A$ an ideal; then the quotient ring $B = A/I$ is Noetherian.

(ii) Let A be a Noetherian integral domain, and $A \subset K$ its field of fractions; let $0 \notin S \subset A$ be a subset, and set

$$B = A[S^{-1}] = \{a/b \in K \mid a \in A, \text{ and } b = 1 \text{ or a product of elements of } S\}.$$

Then B is again Noetherian.

**Proof.** Exercise: in either case the ideals of A can be described in terms of certain ideals of B; see Ex. 3.4 for hints.

**(3.3) Theorem** (Hilbert Basis Theorem). For a ring A,

$$A \text{ Noetherian} \Rightarrow A[X] \text{ Noetherian}.$$

**Proof.** Let $J \subset A[X]$ be any ideal; I prove that J is finitely generated. Define the ideal of leading terms of degree n in J to be

$$J_n = \{a \in A \mid \exists f = aX^n + b_{n-1}X^{n-1} + .. b_0 \in I\}.$$

Then $J_n$ is an ideal of A and $J_n \subset J_{n+1}$ (please provide your own proofs). Hence, using the a.c.c., there exists N such that

$$J_N = J_{N+1} = ..$$

Now build a set of generators of J as follows: for $i \le N$, let $a_{i1},.. a_{im(i)}$ be generators of $J_i$ and, as in the definition of $J_i$, for each of the $a_{ik}$, let $f_{ik} = a_{ik}X^i + .. \in I$ be an element of degree i and leading term $a_{ik}$.

I claim that the set

$$\{f_{ik} \mid i = 0,.. N, k = 1,.. m(i)\}$$

just constructed generates J: for given $g \in J$, suppose $\deg g = m$. Then the leading term of g is $bX^m$ with $b \in J_m$, so that by what I know about $J_m$, I can write $b = \sum c_{m'k}a_{m'k}$ (here $m' = m$ if $m \le N$, otherwise $m' = N$). Then consider $g_1 = g - \sum c_{m'k}(X^{m-m'})f_{m'k}$: by construction the term of degree m is zero, so that $\deg g_1 \le \deg g - 1$; by induction, I can therefore write out g as a combination of $f_{ik}$, so that these generate J. Q.E.D.

**Corollary.** If k is a field, then a finitely generated k-algebra is Noetherian.

A finitely generated k-algebra is a ring of the form $A = k[a_1,.. a_n]$, so that A is generated as a ring by k and $a_1,.. a_n$; clearly, every such ring is isomorphic to a quotient of the polynomial ring, $A \cong k[X_1,.. X_n]/I$. A field is Noetherian, and by induction on (3.3), $k[X_1,.. X_n]$ is Noetherian; finally, passing to the quotient is OK by (3.2). Q.E.D.

**(3.4) The correspondence V.** $k$ is any field, and $A = k[X_1,.. X_n]$. Following an almost universal idiosyncracy of algebraic geometers [1)], I write $\mathbb{A}^n{}_k = k^n$ for the n-dimensional affine space over $k$; given a polynomial $f(X_1,.. X_n) \in A$ and a point $P = (a_1,.. a_n) \in \mathbb{A}^n{}_k$, the element $f(a_1,.. a_n) \in k$ is thought of as 'evaluating the function $f$ at $P$'. Define a correspondence

$$\{\text{ideals } J \subset A\} \xrightarrow{V} \{\text{subsets } X \subset \mathbb{A}^n{}_k\}$$

by

$$J \mapsto V(J) = \{P \in \mathbb{A}^n{}_k \mid f(P) = 0 \text{ for all } f \in J\}.$$

**Definition.** A subset $X \subset \mathbb{A}^n{}_k$ is an *algebraic set* if $X = V(I)$ for some $I$. (This is the same thing as a variety, but I want to reserve the word.) Notice that by Corollary 3.3, $I$ is finitely generated. If $I = (f_1,.. f_r)$ then clearly

$$V(I) = \{P \in \mathbb{A}^n{}_k \mid f_i(P) = 0 \text{ for } i = 1,.. r\},$$

so that an algebraic set is just a locus of points satisfying a finite number of polynomial equations.

If $I = (f)$ is a principal ideal, then I usually write $V(f)$ for $V(I)$; this is of course the same thing as $V\colon (f=0)$ in the notation of §§1–2.

**(3.5) Proposition-Definition.** The correspondence $V$ satisfies the following formal properties:

(i) $V(0) = \mathbb{A}^n{}_k$; $V(A) = \emptyset$;

(ii) $I \subset J \Rightarrow V(I) \supset V(J)$;

(iii) $V(I_1 \cap I_2) = V(I_1) \cup V(I_2)$;

(iv) $V(\sum_{\lambda\in\Lambda} I_\lambda) = \bigcap_{\lambda\in\Lambda} V(I_\lambda)$.

Hence the algebraic subsets of $\mathbb{A}^n{}_k$ form the closed sets of a topology on $\mathbb{A}^n{}_k$, the *Zariski topology*.

The above properties are quite trivial, with the exception of the inclusion $\subset$ in (iii). For this, suppose $P \notin V(I_1) \cup V(I_2)$; then there exist $f \in I_1$, $g \in I_2$ such that $f(P) \neq 0$, $g(P) \neq 0$. So $fg \in I_1 \cap I_2$, but $fg(P) \neq 0$, and therefore $P \notin V(I_1 \cap I_2)$. Q.E.D.

---

1) $\mathbb{A}^n$ is thought of as a variety, whereas $k^n$ is just a point set. Think of this as pure pedantry if you like; compare (4.6) below, as well as (8.3).

The Zariski topology on $\mathbb{A}^n_k$ induces a topology on any algebraic set $X \subset \mathbb{A}^n_k$: the closed subsets of X are the algebraic subsets.

It's important to notice that the Zariski topology on a variety is very weak, and is quite different from the familiar topology of metric spaces like $\mathbb{R}^n$. As an example, a Zariski closed subset of $\mathbb{A}^1_k$ is either the whole of $\mathbb{A}^1_k$ or is finite; see Ex. 3.12 for a description of the Zariski topology on $\mathbb{A}^2_k$. If $k = \mathbb{R}$ or $\mathbb{C}$ then Zariski closed sets are also closed for the ordinary topology, since polynomial functions are continuous. In fact they're very special open or closed subsets: a non-empty Zariski open subset of $\mathbb{R}^n$ is the complement of a subvariety, so it is automatically dense in $\mathbb{R}^n$.

The Zariski topology may cause trouble to some students; since it is only being used as a language, and has almost no content, the difficulty is likely to be psychological rather than technical.

**(3.6) The correspondence I.** As a kind of inverse to V there is a correspondence

$$\{\text{ideals } J \subset A\} \xleftarrow{\ I\ } \{\text{subsets } X \subset \mathbb{A}^n_k\}$$

by

$$I(X) = \{f \in A \mid f(P) = 0 \text{ for all } P \in X\} \longleftarrow\!\!\mid \quad X.$$

That is, I takes a subset X to the ideal of functions vanishing on it.

**Proposition.** (a) $X \subset Y \Rightarrow I(X) \supset I(Y)$;

(b) for any subset $X \subset \mathbb{A}^n_k$, I have $X \subset V(I(X))$, with equality if and only if X is an algebraic set;

(c) for $J \subset A$, I have $J \subset I(V(J))$; the inclusion may well be strict.

**Proof.** (a) is trivial. The two inclusion signs in (b) and (c) are tautologous: if I(X) is defined as the set of functions vanishing at all points of X, then for any point of X, all the functions of I(X) vanish at it. And indeed conversely, if not more so, just as I was about to say myself, Piglet.

The remaining part of (b) is easy: if $X = V(I(X))$ then V is certainly an algebraic set, since it's of the form V(ideal). Conversely, if $X = V(I_0)$ is an algebraic set, then I(X) contains at least $I_0$, so $V(I(X)) \subset V(I_0) = X$.

There are two different ways in which the inclusion $J \subset I(V(J))$ in (c) may be strict. It's most important to understand these, since they lead directly to the correct statement of the Nullstellensatz.

**Example 1.** Suppose that the field $k$ is not algebraically closed, and let $f \in k[X]$ be a non-constant polynomial not having a root in $k$. Consider the ideal $J = (f) \subset k[X]$. Then $J \neq k[X]$, since $1 \notin J$. But

$$V(J) = \{P \in \mathbb{A}^1_k \mid f(P) = 0\} = \emptyset.$$

Therefore $I(V(J)) = k[X]$ (since any function vanishes at all points of the empty set).

So if your field is not algebraically closed, you may not get enough zeros. A rather similar example: in $\mathbb{R}^2$, the polynomial $X^2 + Y^2$ defines the single point $P = (0, 0)$, so $V(X^2 + Y^2) = \{P\}$. But then many more polynomials vanish on $\{P\}$ than just the multiples of $X^2 + Y^2$, and in fact $I(P) = (X, Y)$.

**Example 2.** For any $f \in k[X_1,.. X_n]$ and $a \geq 2$, $f^a$ defines the same locus as $f$, that is $f^a(P) = 0 \iff f(P) = 0$. So $V(f^a) = V(f)$, and $f \in I(V(f^a))$, but usually $f \notin (f^a)$. The trouble here is already present in $\mathbb{R}^2$: in §1, mention was made of the 'double line' defined by $X^2 = 0$. The only meaning that can be attached to this is the line $(X = 0)$ deemed to have multiplicity $2$; but the point set itself doesn't understand that it's being deemed.

(3.7) **Irreducible algebraic set.** An algebraic set $X \subset \mathbb{A}^n_k$ is *irreducible* if there does not exist a decomposition

$$X = X_1 \cup X_2 \quad \text{with} \quad X_1, X_2 \subsetneq X$$

of $X$ as a union of two strict algebraic subsets. For example, the algebraic subset $V(xy) \subset \mathbb{A}^2_k$ is the locus consisting of the two coordinate axes, and is obviously the union of $V(x)$ and $V(y)$, hence reducible.

**Proposition.** (a) Let $X \subset \mathbb{A}^n_k$ be an algebraic set and $I(X)$ the corresponding ideal; then

$$X \text{ is irreducible} \iff I(X) \text{ is prime.}$$

(b) Any algebraic set $X$ has a (unique) expression

$$X = X_1 \cup .. X_r \qquad (*)$$

with $X_i$ irreducible and $X_i \not\subset X_j$ for $i \neq j$.

The $X_i$ in $(*)$ are the *irreducible components* of $X$.

**Proof.** (a) In fact I prove that $X$ is reducible $\iff$ $I(X)$ is not prime.

($\Rightarrow$) Suppose $X = X_1 \cup X_2$ with $X_1, X_2 \subsetneq X$ algebraic subsets. Then $X_1 \subsetneq X$ means that there exists $f_1 \in I(X_1) \setminus I(X)$, and similarly $X_2 \subsetneq X$ gives

$f_2 \in I(X_2) \setminus I(X)$. Now the product $f_1f_2$ vanishes at all points of $X$, so $f_1f_2 \in I(X)$. Therefore $I(X)$ is not prime.

($\Leftarrow$) Suppose that $I(X)$ is not prime; then there exist $f_1, f_2 \notin I(X)$ such that $f_1f_2 \in I(X)$. Let $I_1 = (I(X), f_1)$ and $V(I_1) = X_1$; then $X_1 \subsetneq X$ is an algebraic subset; similarly, setting $I_2 = (I(X), f_2)$ and $V(I_2) = X_2$ gives $X_2 \subsetneq X$. But $X \subset X_1 \cup X_2$, since for all $P \in X$, $f_1f_2(P) = 0$ implies that either $f_1(P) = 0$ or $f_2(P) = 0$.

(b) First of all, I establish the following proposition: the algebraic subsets of $\mathbb{A}^n_k$ satisfy the descending chain condition, that is, every chain

$$X_1 \supset X_2 \supset .. \; X_n \supset ..$$

eventually stops with $X_N = X_{N+1} = \ldots$. This is because

$$I(X_1) \subset I(X_2) \subset .. \; I(X_n) \subset ..$$

is an ascending chain of ideals of $A$, and this stops, giving $X_N = X_{N+1} = \ldots$. Thus just as in (3.1),

$$\text{any non-empty set } \Sigma \text{ of algebraic subsets of } \mathbb{A}^n_k \text{ has a minimal element.} \qquad (!)$$

Now to prove (b), let $\Sigma$ be the set of algebraic subsets of $\mathbb{A}^n_k$ which do not have a decomposition $(*)$. If $\Sigma = \varnothing$ then (b) is proved. On the other hand, if $\Sigma \neq \varnothing$ then by (!), there must be a minimal element $X \in \Sigma$, and this leads speedily to one of two contradictions: if $X$ is irreducible, then $X \notin \Sigma$, a contradiction; if $X$ is reducible, then $X = X_1 \cup X_2$, with $X_1, X_2 \subsetneq X$, so that by minimality of $X \in \Sigma$, I get $X_1, X_2 \notin \Sigma$. So each of $X_1, X_2$ has a decomposition $(*)$ as a union of irreducibles, and putting them together gives a decomposition for $(*)$, so $X \notin \Sigma$. This contradiction proves $\Sigma = \varnothing$. This proves the existence part of (b). The uniqueness is an easy exercise, see Ex. 3.8. Q.E.D.

The proof of (b) is a typical algebraist's proof: it's logically very neat, but almost completely hides the content: the real point is that if $X$ is not irreducible, then it breaks up as $X = X_1 \cup X_2$, and then you ask the same thing about $X_1$ and $X_2$, and so on; eventually, you must get to irreducible algebraic sets, since otherwise you'd get an infinite descending chain.

**(3.8)** I now want to state and prove the Nullstellensatz. There is an intrinsic difficulty in any proof of the Nullstellensatz, and I choose to break it up into two segments. Firstly I state without proof an assertion in commutative algebra, which will be proved in (3.15) below (in fact parts of the proof will have strong geometric content).

**Hard Fact.** Let $k$ be a (infinite) field, and $A = k[a_1,.. a_n]$ a finitely generated $k$-algebra. Then

$$A \text{ is a field } \Rightarrow A \text{ is algebraic over } k.$$

Just to give a rough idea why this is true, notice that if $t \in A$ is transcendental over $k$, then $k[t]$ is a polynomial ring, so *has infinitely many primes* (by Euclid's argument). Hence the extension $k \subset k(t)$ is not finitely generated as $k$-algebra: finitely many elements $p_i/q_i \in k(t)$ can have only finitely many primes among their denominators.

**(3.9) Definition.** If $I$ is an ideal of $A$, the *radical* of $I$ is

$$\operatorname{rad} I = \sqrt{I} = \{f \in A \mid f^n \in I \text{ for some } n\}.$$

$\operatorname{rad} I$ is an ideal, since $f, g \in \operatorname{rad} I \Rightarrow f^n, g^m \in I$ for suitable $n, m$, and therefore

$$(f+g)^r = \sum \binom{n}{r} f^a g^{r-a} \in I \quad \text{if} \quad r \geq n+m-1.$$

An ideal $I$ is *radical* if $I = \operatorname{rad} I$.

Note that a prime ideal is radical. It's not hard to see that in a UFD like $k[X_1,.. X_n]$, a principal ideal $I = (f)$ where $f = \prod f_i^{n(i)}$ (factorisation into distinct prime factors), has $\operatorname{rad} I = (f_{red})$, where $f_{red} = \prod f_i$.

**(3.10) Nullstellensatz** (Hilbert's zeros theorem). Let $k$ be an algebraically closed field.

(a) Every maximal ideal of the polynomial ring $A = k[X_1,.. X_n]$ is of the form $m_P = (X_1 - a_1,.. X_n - a_n)$ for some point $P = (a_1,.. a_n) \in \mathbb{A}^n_k$; that is, it's the ideal $I(P)$ of all functions vanishing at $P$.

(b) Let $J \subsetneq A$ be an ideal, $J \neq (1)$; then $V(J) \neq \emptyset$.

(c) For any $J \subset A$,

$$I(V(J)) = \operatorname{rad} J.$$

The essential content of the theorem is (b), which says that if an ideal $J$ is not the whole of $k[X_1,.. X_n]$, then it will have zeros in $\mathbb{A}^n_k$. Note that (b) is completely false if $k$ is not algebraically closed, since if $f \in k[X]$ is a non-constant polynomial then it will not generate the whole of $k[X]$ as an ideal, but $V(f) =$

$\emptyset \subset \mathbb{A}^1_k$ is perfectly possible. The name of the theorem (*Nullstelle* = zero of a polynomial + *Satz* = theorem) should help to remind you of the content (but stick to the German if you don't want to be considered an ignorant peasant).

**Corollary**. The correspondences V and I

$$\{\text{ideals } I \subset A\} \xleftrightarrow{V,\, I} \{\text{subsets } X \subset \mathbb{A}^n_k\}$$

induce bijections

$$\cup \qquad\qquad \cup$$

$$\{\text{radical ideals}\} \longleftrightarrow \{\text{algebraic subsets}\}$$

and

$$\cup \qquad\qquad \cup$$

$$\{\text{prime ideals}\} \longleftrightarrow \{\text{irreducible alg. subsets}\}.$$

This holds because $V(I(X)) = X$ for any algebraic set $X$ by (3.6), (b), and $I(V(J)) = J$ for any radical ideal $J$ by (c) above.

**Proof of NSS** (assuming (3.8)). (a) Let $m \subset k[X_1,.. X_n]$ be a maximal ideal; write $K = k[X_1,.. X_n]/m$, and $\varphi$ for the composite of natural maps $\varphi\colon k \longrightarrow k[X_1,.. X_n] \to K$. Then $K$ is a field (since $m$ is maximal), and it is finitely generated as k-algebra (since its generated by the images of the $X_i$). So by (3.8), $\varphi\colon k \to K$ is an algebraic field extension. But $k$ is algebraically closed, hence $\varphi$ is an isomorphism.

Now for each i, $X_i \in k[X_1,.. X_n]$ maps to some element $b_i \in K$; so taking $a_i = \varphi^{-1}(b_i)$ gives $X_i - a_i \in \operatorname{Ker}\{k[X_1,.. X_n] \to K\} = m$. Hence there exist $a_1,.. a_n \in k$ such that $(X_1 - a_1,.. X_n - a_n) \subset m$. On the other hand, it's clear that the left-hand side is a maximal ideal, so $(X_1 - a_1,.. X_n - a_n) = m$. This proves (a).

(a) $\Longrightarrow$ (b) This is easy. If $J \neq A = k[X_1,.. X_n]$ then there exists a maximal ideal $m$ of $A$ such that $J \subset m$ (the existence of $m$ is easy to check, using the a.c.c.). By (a), $m$ is of the form $m = (X_1 - a_1,.. X_n - a_n)$; then $J \subset m$ just means that $f(P) = 0$ for all $f \in J$, where $P = (a_1,.. a_n)$. Therefore $P \in V(J)$.

(b) $\Longrightarrow$ (c) This requires a cunning trick. Let $J \subset k[X_1,.. X_n]$ be any ideal, and $f \in k[X_1,.. X_n]$. Introduce another variable Y, and consider the new ideal

$$J_1 = (J, fY - 1) \subset k[X_1,.. X_n, Y]$$

generated by J and $fY - 1$. Roughly speaking, $V(J_1)$ is the variety consisting of $P \in V(J)$ such that $f(P) \neq 0$. More precisely, a point $Q \in V(J_1) \subset \mathbb{A}^{n+1}_k$ is an $(n + 1)$-tuple $Q = (a_1,.. a_n, b)$ such that

$$g(a_1,.. a_n) = 0 \text{ for all } g \in J, \text{ that is, } P = (a_1,.. a_n) \in V(J),$$

and

$$f(P)\cdot b = 1, \text{ that is } f(P) \neq 0 \text{ and } b = f(P)^{-1}.$$

Now suppose that $f(P) = 0$ for all $P \in V(J)$; then clearly, from what I've just said, $V(J_1) = \varnothing$. So I can use (b) to deduce that $1 \in J_1$, that is, there exists an expression

$$1 = \sum g_i f_i + g_0(fY - 1) \in k[X_1,.. X_n, Y] \qquad (**)$$

with $f_i \in J$, and $g_0, g_i \in k[X_1,.. X_n, Y]$.

Consider the way in which $Y$ appears in the right-hand side of (**): apart from its explicit appearance in the second term, it can appear in each of the $g_i$; suppose that $Y^N$ is the highest power of $Y$ appearing in any of $g_0, g_i$. If I then multiply through both sides of (**) by $f^N$, I get a relation of the form

$$f^N = \sum G_i(X_1,.. X_n, fY)f_i + G_0(X_1,.. X_n, fY)(fY - 1); \qquad (***)$$

here $G_i$ is just $f^N g_i$ written out as a polynomial in $X_1,.. X_n$ and $(fY)$.

(***) is just an equality of polynomials in $k[X_1,.. X_N, Y]$, so I can reduce it modulo $(fY - 1)$ to get

$$f^N = \sum h_i(X_1,.. X_n)f_i \in k[X_1,.. X_N, Y]/(fY - 1);$$

both sides of the equation are elements of $k[X_1,.. X_n]$. Since the natural homomorphism $k[X_1,.. X_n] \hookrightarrow k[X_1,.. X_N, Y]/(fY - 1)$ is injective (it is just the inclusion of $k[X_1,.. X_n]$ into $k[X_1,.. X_n][f^{-1}]$, as a subring of its field of fractions), it follows that

$$f^N = \sum h_i(X_1,.. X_n)f_i \in k[X_1,.. X_N];$$

that is, $f^N \in J$ for some $N$. Q.E.D.

**Remark.** Several of the textbooks cut the argument short by just saying that (**) is an identity, so it remains true if we set $Y = f^{-1}$. This is of course perfectly valid, but I have preferred to spell it out in detail.

**(3.11) Worked examples.** (a) Hypersurfaces. The simplest example of a variety is the hypersurface $V(f)\colon (f = 0) \subset \mathbb{A}^n_k$. If $k$ is algebraically closed, there is just the obvious correspondence between irreducible elements $f \in k[X_1,.. X_n]$ and irreducible hypersurfaces: it follows from the Nullstellensatz that two distinct irreducible polynomials $f_1, f_2$ (not multiples of one another) define different hypersurfaces $V(f_1)$ and $V(f_2)$. This is not at all obvious (for example, it's false over $\mathbb{R}$), although it can be proved without using the Nullstellensatz by *elimination theory*, a much more explicit method with a nice 19th century flavour; see Ex. 3.13.

(b) Once past the hypersurfaces, most varieties are given by 'lots' of equations; contrary to intuition, it is usually the case that the ideal $I(X)$ needs many generators, that is, many more than the codimension of $X$. I give an example of a curve $C \subset \mathbb{A}^3_k$ for which $I(C)$ needs 3 generators; assume that $k$ is an infinite field.

Consider first $J = (uw - v^2, u^3 - vw)$. Then $J$ is not prime, since

$$J \ni w(uw - v^2) - v(u^3 - vw) = u(w^2 - u^2v),$$

but $u, w^2 - u^2v \notin J$. Therefore

$$V(J) = V(J, u) \cup V(J, w^2 - u^2v);$$

obviously, $V(J, u) = \text{line } (u = v = 0)$. I claim that the other component $C = V(J, w^2 - u^2v)$ is an irreducible curve; indeed, $C$ is given by

$$uw = v^2, \; u^3 = vw, \; w^2 = u^2v.$$

I claim that $C \subset \mathbb{A}^3$ is the image of the map $\varphi\colon \mathbb{A}^1 \to C \subset \mathbb{A}^3$ given by $t \mapsto t^3, t^4, t^5$: to see this, if $u \neq 0$ then $v, w \neq 0$. Set $t = v/u$, then $t = w/v$ and $t^2 = (v/u)(w/v) = w/u$. Hence $v = w^2/u^2 = t^4$, $u = v/(v/u) = t^4/t = t^3$, and $w = tv = t^5$. Now $C$ is irreducible, since if $C = X_1 \cup X_2$ with $X_i \subset C$, and $f_i(u, v, w) \in I(X_i)$, then for all $t$, one of $f_i(t^3, t^4, t^5)$ must vanish. Since a non-zero polynomial has at most a finite number of zeros, one of $f_1, f_2$ must vanish identically, so $f_i \in I(C)$.

This example is of a nice 'monomial' kind; in general it might be quite tricky to guess the irreducible components of a variety, and even more so to prove that they are irreducible. A similar example is given in Ex. 3.11.

**(3.12) Finite algebras.** I now start on the proof of (3.8). Let $A \subset B$ be rings. As usual, $B$ is said to be *finitely generated* over $A$ (or f.g. as $A$-algebra) if there exist finitely many elements $b_1,.. b_n$ such that $B = A[b_1,..b_n]$, so that $B$ is generated as a ring by $A$ and $b_1,.. b_n$.

Contrast with the following definition: $B$ is a *finite* $A$-*algebra* if there exist finitely many elements $b_1,.. b_n$ such that $B = Ab_1 + .. Ab_n$, that is, $B$ is finitely generated as $A$-module. The crucial distinction here is between generation as ring (when you're allowed any polynomial expressions in the $b_i$), and as module (the $b_i$ can only occur linearly). For example, $k[X]$ is a finitely generated $k$-algebra (it's generated by 1 element $X$), but is not a finite $k$-algebra (since it has infinite dimension as $k$-vector space).

**Proposition.** (i) Let $A \subset B \subset C$ be rings; then

$$B \text{ a finite } A\text{-algebra and } C \text{ a finite } B\text{-algebra} \Rightarrow C \text{ a finite } A\text{-algebra.}$$

(ii) If $A \subset B$ is a finite $A$-algebra and $x \in B$ then $x$ satisfies a monic equation over $A$, that is, there exists a relation

$$x^n + a_{n-1}x^{n-1} + .. a_0 = 0 \quad \text{with} \quad a_i \in A$$

(note that the leading coefficient is 1).

(iii) Conversely, if $x$ satisfies a monic equation over $A$, then $B = A[x]$ is a finite $A$-algebra.

**Proof**. (i) and (iii) are easy exercises (compare similar results for field extensions). For (ii), I use a rather non-obvious 'determinant trick' (which I didn't think of for myself): suppose $B = \sum Ab_i$; for each $i$, $xb_i \in B$, so there exist constants $a_{ij} \in A$ such that

$$xb_i = \textstyle\sum_j a_{ij}b_j.$$

This can be written

$$\textstyle\sum_j (x\delta_{ij} - a_{ij})b_j = 0,$$

where $\delta_{ij}$ is the identity matrix. Now let $M$ be the matrix with

$$M_{ij} = (x\delta_{ij} - a_{ij}),$$

and set $\Delta = \det M$. Then by standard linear algebra, (writing $\mathbf{b}$ for the column vector with entries $(b_1,.. b_n)$),

$$M\mathbf{b} = 0, \quad \text{hence} \quad 0 = (M^{adj})M\mathbf{b} = \Delta\mathbf{b},$$

(where $(M^{adj})$ is the adjoint matrix), and therefore $\Delta b_i = 0$ for all $i$. However, $1_B \in B$ is a linear combination of the $b_i$, so that $\Delta = \Delta \cdot 1_B = 0$, and I've won my relation: $\det(x\delta_{ij} - a_{ij}) = 0$. This is obviously a monic relation for $x$ with coefficients in $A$. Q.E.D.

**(3.13) Noether normalisation.**

**Theorem** (Noether normalisation lemma). Let $k$ be an infinite field, and $A = k[a_1,.. a_n]$ a finitely generated $k$-algebra. Then there exist $m \leq n$ and $y_1,.. y_m \in A$ such that

(i) $y_1,.. y_m$ are algebraically independent over $k$;

and

(ii) $A$ is a finite $k[y_1,.. y_m]$-algebra.

((i) means as usual that there are no non-zero polynomial relations holding between the $y_i$; an algebraist's way of saying this is that the natural (surjective) map $k[Y_1,.. Y_m] \rightarrow k[y_1,.. y_m] \subset A$ is injective.)

It is being asserted that, as you might expect, the extension of rings can be built up by first throwing in algebraically independent elements, then 'making an algebraic extension'; however, the statement (ii) is far more precise than this, since it says that every element of $A$ is not just algebraic over $k[y_1,.. y_m]$, but satisfies a *monic* equation over it.

**Proof.** Let $I$ be the kernel of the natural surjection,

$$I = \ker \{k[X_1,.. X_n] \rightarrow k[a_1,.. a_n] = A\}.$$

Suppose that $0 \neq f \in I$; the idea of the proof is to replace $X_1,.. X_{n-1}$ by certain $X'_1.. X'_{n-1}$ so that $f$ becomes a monic equation for $a_n$ over $A' = k[a'_1,.. a'_{n-1}]$. So write

$$a'_1 = a_1 - \alpha_1 a_n$$

$$..$$

$$a'_{n-1} = a_{n-1} - \alpha_{n-1} x_n$$

(where the $\alpha_i$ are elements of $k$ to be specified later). Then

$$0 = f(a'_1 + \alpha_1 a_n,.. a'_{n-1} + \alpha_{n-1} a_n, a_n).$$

**Claim.** For suitable choice of $\alpha_1,.. \alpha_{n-1} \in k$, the polynomial

$$f(X'_1 + \alpha_1 X_n,.. X'_{n-1} + \alpha_{n-1} X_n, X_n)$$

is monic in $X_n$.

Using the claim, the theorem is proved by induction on n: if $I = 0$ then there's nothing to prove, since $a_1,.. a_n$ are algebraically independent. Otherwise, pick $0 \neq f \in I$, and let $\alpha_1,.. \alpha_{n-1}$ be as in the claim; then f gives a monic relation satisfied by $a_n$ with coefficients in $A' = k[a'_1,.. a'_{n-1}] \subset A$. By the inductive assumption, there exist $y_1,.. y_m \in A'$ such that

(1) $y_1,.. y_m$ are algebraically independent over k;

(2) $A'$ is a finite $k[y_1,.. y_m]$-algebra.

Then $A = A'[a_n]$ is finite over $A'$ (by (3.11, iii)), so by (3.11, i), A is finite over $k[y_1,.. y_m]$, proving the theorem.

It only remains to prove the claim. Let $d = \deg f$, and write

$$f = F_d + G,$$

with $F_d$ homogeneous of degree d, and $\deg G \leq d - 1$. Then

$$f(X_1,.. X_{n-1}, X_n) = f(X'_1 + \alpha_1 X_n,.. X'_{n-1} + \alpha_{n-1}X_n, X_n)$$

$$= F_d(\alpha_1,.. \alpha_{n-1}, 1)\cdot X_n^d + (\text{stuff involving } X_n \text{ to power } \leq d - 1);$$

I'm now home provided that $F_d(\alpha_1,.. \alpha_{n-1}, 1) \neq 0$. Since $F_d$ is a non-zero polynomial, it's not hard to check that this is the case for 'almost all' values of $\alpha_1,.. \alpha_{n-1}$ (the proof of this is discussed in Ex. 3.13). Q.E.D.

**(3.14) Remarks.** (I) In fact, the proof of (3.13) shows that $y_1,.. y_m$ can be chosen to be m general linear forms in $a_1,.. a_n$. To understand the significance of (3.13), write $I = \ker \{k[X_1,.. X_n] \to k[a_1,.. a_n] = A\}$, and assume for simplicity that I is prime. Consider $V = V(I) \subset \mathbb{A}^n_k$; let $\pi: \mathbb{A}^n_k \to \mathbb{A}^m_k$ be the linear projection defined by $y_1,.. y_m$, and $p = \pi|_V: V \to \mathbb{A}^m_k$. It can be seen that the conclusions (i) and (ii) of (3.13) imply that above every $P \in \mathbb{A}^m_k$, $p^{-1}(P)$ is a finite non-empty set (see Ex. 3.16).

(II) The proof of (3.13) has also a simple geometric interpretation: choosing n-1 linear forms in the n variables $X_1,.. X_n$ corresponds to making a linear projection $\pi: \mathbb{A}^n_k \to \mathbb{A}^{n-1}_k$; the fibres of $\pi$ then form a (n-1)-dimensional family of parallel lines. Having chosen the polynomial $f \in I$, it is not hard to see that f gives rise to a monic relation in the final $X_n$ if and only if none of the parallel lines are asymptotes of the variety (f = 0); in terms of projective geometry, this means that the point at infinity $(0, \alpha_1,.. \alpha_{n-1}, 1) \in \mathbb{P}^{n-1}_k$ specifying the parallel projection does not belong to the projective closure of (f = 0).

(III) The above proof of (3.13) does not work for a finite field (see Ex. 3.14). However, the theorem itself is true without any condition on k (see

[Mumford, Introduction, p.4] or [Atiyah and Macdonald, (7.9)]).

**(3.15) Proof of (3.8).** Let $A = k[a_1,.. a_n]$ be a finitely generated k-algebra. Suppose $y_1,.. y_m \in A$ are as in (3.13), and write $B = k[y_1,.. y_m]$. Then A is a finite B-algebra, and it is given that A is a field. If I knew that B is a field, it would follow at once that $m = 0$, so that A is a finite k-algebra, that is, a finite field extension of k, and (3.8) would be proved. Therefore it remains only to prove the following statement:

**Lemma.** If A is a field, and $B \subset A$ is a subring such that A is a finite B-algebra, then B is a field.

**Proof.** For any $0 \neq b \in B$, the inverse $b^{-1} \in A$ exists in A. Now by (3.12, ii), the finiteness implies that $b^{-1}$ satisfies a monic equation over B, that is, there exists a relation

$$b^{-n} + a_{n-1}b^{-(n-1)} + .. a_1 b^{-1} + a_0 = 0, \quad \text{with} \quad a_i \in B;$$

then multiplying through by $b^{n-1}$,

$$b^{-1} = -(a_{n-1} + a_{n-2}b + .. a_0 b^{n-1}) \in B.$$

Therefore B is a field. This proves (3.8) and completes the proof of NSS.

**(3.16)** For the purposes of arranging that everything goes through in characteristic p, it is useful to add a tiny precision. I'm only going to use this in one place in the sequel, so if you can't remember too much about separability from Galois theory, don't lose too much sleep over it (GOTO (3.17)).

**Addendum.** Under the conditions of (3.14), if furthermore k is algebraically closed, and A is an integral domain with field of fractions K then $y_1,.. y_m \in A$ can be chosen as above so that (i) and (ii) hold, and in addition

(iii) $k(y_1,.. y_m) \subset K$ is a separable extension.

**Proof.** If k is of characteristic 0, then every field extension is separable; suppose therefore that k has characteristic p. Since A is an integral domain, I is prime; hence if $I \neq 0$, it contains an irreducible element f. Now for each i, there is a dichotomy: either f is separable in $X_i$, or $f \in k[X_1,.. X_i^p,.. X_n]$.

**Claim.** If f is inseparable in each $X_i$, then $f = g^p$ for some g, contradicting the irreducibility of f.

The assumption is that f is of the form:

$$f = F(X_1{}^p,.. X_n{}^p), \text{ with } F \in k[X_1,.. X_n].$$

If this happens, let $g \in k[X_1,.. X_n]$ be the polynomial obtained by taking the pth root of each coefficient of F; then making repeated use of the standard identity $(a + b)^p = a^p + b^p$ in characteristic p, it is easy to see that $f = g^p$.

It follows that any irreducible f is separable in at least one of the $X_i$, say in $X_n$. Then arguing exactly as above,

$$f(X'_1 + \alpha_1 X_n,.. X'_{n-1} + \alpha_{n-1}X_n, X_n)$$

provides a monic, separable relation for $a_n$ over $A' = k[a'_1,.. a'_n]$. The result then follows by the same induction argument, using this time the fact that a composite of separable field extensions is separable. Q.E.D.

**(3.17) Reduction to a hypersurface.** Recall the following result from Galois theory:

**Primitive element theorem.** Let K be an infinite field, and $K \subset L$ a finite separable field extension; then there exists $x \in L$ such that $L = K(x)$. Moreover, if L is generated over K by elements $z_1,.. z_k$, the element x can be chosen to be a linear combination $\sum_i \alpha_i z_i$.

(This follows at once from the Fundamental Theorem of Galois theory: if $K \subset M$ is the normal closure of L over K then $K \subset M$ is a finite Galois field extension, so that by the Fundamental Theorem there only exist finitely many intermediate field extensions between K and M. The intermediate subfields between K and L form a finite collection $\{K_j\}$ of K-vector subspaces of L, so that I can choose $x \in L$ not belonging to any of these. If $z_1,.. z_k$ are given, not all belonging to any $K_i$, then x can be chosen as a K-linear combination of the $z_i$. Then $K(x) = L$.)

**Corollary.** Under the hypotheses of the Noether normalisation lemma (3.13), there exist $y_1,.. y_{m+1} \in A$ such that $y_1,.. y_m$ satisfy the conclusion of (3.13), and in addition, the field of fractions K of A is generated over k by $y_1,.. y_{m+1}$.

**Proof.** According to (3.17), I can arrange that K is a separable extension of $k(y_1,.. y_m)$. If $A = k[x_1,.. x_n]$, then the $x_i$ certainly generate K as a field extension of $k(y_1,.. y_m)$, so that a suitable linear combination $y_{m+1}$ of the $x_i$ with coefficients in $k(y_1,.. y_m)$ generates the field extension; clearing denominators,

$y_{m+1}$ can be taken as a linear combination of the $x_i$ with coefficients in $k[y_1,.. y_m]$, hence as an element of A. Q.E.D.

Algebraically, what I have proved is that the field extension $k \subset K$, while not necessarily purely transcendental, can be generated as a composite of a purely transcendental extension $k \subset k(y_1,.. y_m) = K_0$ followed by a primitive algebraic extension $K_0 \subset K = K_0(y_{m+1})$. In other words, $K = k(y_1,.. y_{m+1})$, with only one algebraic dependence between the generators. The geometric significance of the result will become clear in (5.10).

## Exercises to §3.

**3.1.** An integral domain A is a *principal ideal domain* if every ideal I of A is principal, that is of the form $I = (a)$; show directly that the ideals in a PID satisfy the a.c.c.

**3.2.** Show that an integral domain A is a UFD if and only if every ascending chain of principal ideals terminates, and every irreducible element of A is prime.

**3.3.** (i) Prove Gauss's lemma: if A is a UFD and $f, g \in A[X]$ are polynomials with coefficients in A, then a prime element of A that is a common factor of the coefficients of the product fg is a common factor of the coefficients of f or g.

(ii) It is proved in undergraduate algebra that if K is a field then K[X] is a UFD. Prove by induction on n that $k[X_1,.. X_n]$ is a UFD; for this you will need to compare factorisations in $k[X_1,.. X_n]$ with factorisations in $k(X_1,.. X_{n-1})[X_n]$, using Gauss's lemma to clear denominators.

**3.4.** Prove Proposition 3.2, (ii): if A is an integral domain with field of fractions K, and if $0 \notin S \subset A$ is a subset, define

$$B = A[S^{-1}] = \{a/b \in K \mid a \in A, \text{ and } b = 1 \text{ or product of } s\text{'s}\};$$

prove that an ideal I of B is completely determined by its intersection with A, and deduce that A Noetherian $\Rightarrow$ B Noetherian.

**3.5.** Let $J = (XY, XZ, YZ) \subset k[X, Y, Z]$; find $V(J) \subset \mathbb{A}^3$; is it irreducible? Is it true that $J = I(V(J))$? Prove that J cannot be generated by 2 elements. Now let $J' = (XY, (X - Y)Z)$; find $V(J')$, and calculate rad J'.

**3.6.** Let $J = (X^2 + Y^2 - 1, Y - 1)$; find $f \in I(V(J)) \setminus J$.

**3.7.** Let $J = (X^2 + Y^2 + Z^2, XY + XZ + YZ)$; identify $V(J)$ and $I(V(J))$.

**3.8.** Prove that the irreducible components of an algebraic set are unique (this was asserted without proof in (3.7, (b)). That is, given two decompositions $V = \bigcup_{i \in I} V_i = \bigcup_{j \in J} W_j$ of V as a union of irreducibles, assumed to be irredundant (that is, $V_i \not\subset V_{i'}$ for $i \neq i'$), prove that the $V_i$ are just a renumbering of the $W_j$.

**3.9.** Let $f = X^2 - Y^2$ and $g = X^3 + XY^2 - Y^3 - X^2Y - X + Y$; find the irreducible components of $V(f, g) \subset \mathbb{A}^2_{\mathbb{C}}$.

**3.10.** If $J = (uw - v^2, w^3 - u^5)$, show that $V(J)$ has two irreducible components, one of which is the curve $C$ of (3.11).

Prove that the same curve $C$ can be defined by two equations, $uw = v^2$ and $u^5 - 2u^2vw + w^3 = 0$. The point here is that the second equation, restricted to the quadric cone ($uw = v^2$) is trying to be a square.

**3.11.** Let $f = v^2 - uw$, $g = u^4 - vw$, $h = w^2 - u^3v$. Identify the variety $V(f, g, h) \subset \mathbb{A}^3$ in the spirit of (3.11). Find out whether $V(f, g)$, $V(f, h)$ and $V(g, h)$ have any other interesting components.

**3.12.** (i) Prove that for any field $k$, an algebraic set in $\mathbb{A}^1_k$ is either finite or the whole of $\mathbb{A}^1_k$. Deduce that the Zariski topology is the cofinite topology.

(ii) Let $k$ be any field, and let $f, g \in k[X, Y]$ be irreducible elements, not multiples of one another. Prove that $V(f, g)$ is finite. (Hint: Write $K = k(X)$; prove first that $f, g$ have no common factors in the PID $K[Y]$. Deduce that there exist $p, q \in K[Y]$ such that $pf + qg = 1$; now by clearing denominators in $p, q$, show that there exists $h \in k[X]$ and $a, b \in k[X, Y]$ such that $h = af + bg$. Hence conclude that there are only finitely many possible values of the $X$-coordinate of points of $V(f, g)$.)

(iii) Prove that any algebraic set $V \subset \mathbb{A}^2_k$ is a finite union of points and curves.

**3.13.** (a) Let $k$ be an infinite field and $f \in k[X_1,.. X_n]$; suppose that $f$ is non-constant, that is, $f \notin k$. Prove that $V(f) \neq \mathbb{A}^n_k$. (Hint: suppose that $f$ involves $X_n$, and consider $f = \sum a_i(X_1,.. X_{n-1})X_n^i$; now use induction on $n$.)

(b) Now suppose that $k$ is algebraically closed, and let $f$ be as in (a). Suppose that $f$ has degree $m$ in $X_n$, and let $a_m(X_1,.. X_{n-1})X_n^m$ be the leading term; show that wherever $a_m \neq 0$, there is a finite non-empty set of points of $V(f)$ corresponding to every value of $(X_1,.. X_{n-1})$. Deduce in particular that if $n \geq 2$ then $V(f)$ is infinite.

(c) Put together the results of (b) and of Ex. 3.12, (iii) to deduce that if the field $k$ is algebraically closed, then distinct irreducible polynomials $f \in k[X, Y]$ define distinct hypersurfaces of $\mathbb{A}^2_k$ (compare (3.11, (a)).

(d) Generalise the result of (c) to $\mathbb{A}^n_k$.

**3.14.** Give an example to show that the proof of Noether normalisation given in (3.13) fails over a finite field $k$. (Hint: find a polynomial $f(X, Y)$ for which $F_d(\alpha, 1) = \alpha^q - \alpha$, so that $F_d(\alpha, 1) = 0$ for all $\alpha \in k$.)

**3.15.** Let $A$ be a ring and $A \subset B$ a finite $A$-algebra. Prove that if $m$ is a maximal ideal of $A$ then $mB \neq B$. (Hint: By contradiction, suppose $B = mB$; if $B = \sum Ab_i$ then for each $i$, $b_i = \sum a_{ij}b_j$ with $a_{ij} \in m$. Now prove that

$$\Delta = \det(\delta_{ij} - a_{ij}) = 0,$$

and conclude that $1_B \in m$, a contradiction. (See also [Atiyah and Macdonald, Prop. 2.4 and Cor. 2.5].)

**3.16.** Let $A = k[a_1,.. a_n]$ be as in the statement of Noether normalisation (3.13), write $I = \ker\{k[X_1,.. X_n] \to k[a_1,.. a_n] = A\}$, and consider $V = V(I)$ in $\mathbb{A}^n_k$; assume for simplicity that I is prime.

Let $Y_1,.. Y_m$ be general linear forms in $X_1,.. X_m$, and write $\pi: \mathbb{A}^n{}_k \to \mathbb{A}^m{}_k$ for the linear projection defined by $Y_1,.. Y_m$; set $p = \pi_{|V}: V \to \mathbb{A}^m{}_k$. Prove that (i) and (ii) of (3.13) imply that above every $P \in \mathbb{A}^m{}_k, p^{-1}(P)$ is a finite set, and non-empty if k is algebraically closed. (Hint: I contains a monic relation for each $X_i$ over $k[Y_1,.. Y_m]$; the finiteness comes easily from this. For the non-emptiness, use Ex. 3.15 to show that for any $P = (b_1,.. b_m) \in \mathbb{A}^m{}_k$, the ideal $J_P = I + (Y_1 - b_1,.. Y_m - b_m) \neq k[X_1,.. X_m]$. Then apply the non-emptiness assertion of the Nullstellensatz.)

# §4. Functions on varieties

In this section I work over a fixed field k; from (4.8, (II)) onwards, k will be assumed to be algebraically closed. The reader who assumes throughout that $k = \mathbb{C}$ will not lose much, and may gain a psychological crutch. I sometimes omit mention of the field k to simplify notation.

**(4.1) Polynomial functions.** Let $V \subset \mathbb{A}^n_k$ be an algebraic set, and I(V) its ideal. Then the quotient ring $k[V] = k[X_1,.. X_n]/I(V)$ is in a natural way a ring of functions on V. In more detail, define a *polynomial function* on V to be a map $f: V \to k$ of the form $P \mapsto F(P)$, with $F \in k[X_1,.. X_n]$; this just means that f is the restriction of a map $F: \mathbb{A}^n \to k$ defined by a polynomial. By definition of I(V), two elements $F, G \in k[X_1,.. X_n]$ define the same function on V if and only if

$$F(P) - G(P) = 0 \quad \text{for all} \quad P \in V,$$

that is, if and only if $F - G \in I(V)$. Thus I define the *coordinate ring* k[V] by

$$k[V] = \{f: V \to k \mid f \text{ is a polynomial function}\} \cong k[X_1,.. X_n]/I(V).$$

This is the smallest ring of functions on V containing the coordinate functions $X_i$ (together with k), so for once the traditional terminology is not too obscure.

**(4.2) k[V] and algebraic subsets of V.** An algebraic set $X \subset \mathbb{A}^n$ is contained in V if and only if $I(X) \supset I(V)$. On the other hand, ideals of $k[X_1,.. X_n]$ containing I(V) are in obvious bijection with ideals of $k[X_1,.. X_n]/I(V)$. (Think about this if it's not obvious to you: the ideal J with $I(V) \subset J \subset k[X_1,.. X_n]$ corresponds to J/I(V); and conversely, an ideal $J_0$ of $k[X_1,.. X_n]/I(V)$ corresponds to its inverse image in $k[X_1,.. X_n]$.)

Hence the I and V correspondences

$$\{\text{ideals } I \subset k[V]\} \xrightarrow{V} \{\text{subsets } X \subset V\}$$

by

$$I \mapsto V(I) = \{P \in V \mid f(P) = 0 \text{ for all } f \in I\}$$

and

$$\{\text{ideals } J \subset k[V]\} \quad \xleftarrow{I} \quad \{\text{subsets } X \subset V\}$$

by

$$I(X) = \{f \in A \mid f(P) = 0 \text{ for all } P \in X\} \quad \longleftarrow\!\!\!\dashv \quad X$$

are defined as in §3, and have similar properties. In particular V has a Zariski topology, in which the closed sets are the algebraic subsets (this is of course the subspace topology of the Zariski topology of $\mathbb{A}^n$).

**Proposition.** Let $V \subset \mathbb{A}^n$ be an algebraic subset. The following conditions are equivalent:

(i) V is irreducible;

(ii) any two open subsets $\emptyset \neq U_1, U_2 \subset V$ have $U_1 \cap U_2 \neq \emptyset$;

(iii) any non-empty open subset $U \subset V$ is dense.

This is all quite trivial: V is irreducible means that V is not a union of two proper closed subsets; (ii) is just a restatement in terms of complements, since

$$U_1 \cap U_2 = \emptyset \iff V = (V - U_1) \cup (V - U_2).$$

A subset of a topological space is dense if and only if it meets every open, so that (iii) is just a restatement of (ii).

**(4.3) Polynomial maps.** Let $V \subset \mathbb{A}^n$ and $W \subset \mathbb{A}^m$ be algebraic sets; write $X_1, .. X_n$ and $Y_1, .. Y_m$ for the coordinates on $\mathbb{A}^n$ and $\mathbb{A}^m$ respectively.

**Definition.** A map $f: V \to W$ is a *polynomial map* if there exist m polynomials $F_1, .. F_m \in k[X_1, .. X_n]$ such that

$$f(P) = (F_1(P), .. F_m(P)) \in \mathbb{A}^m{}_k \quad \text{for all } P \in V.$$

This is an obvious generalisation of the above notion of a polynomial function.

**Claim.** A map $f: V \to W$ is a polynomial map if and only if for all j, the composite map $f_j = Y_j \circ f \in k[V]$:

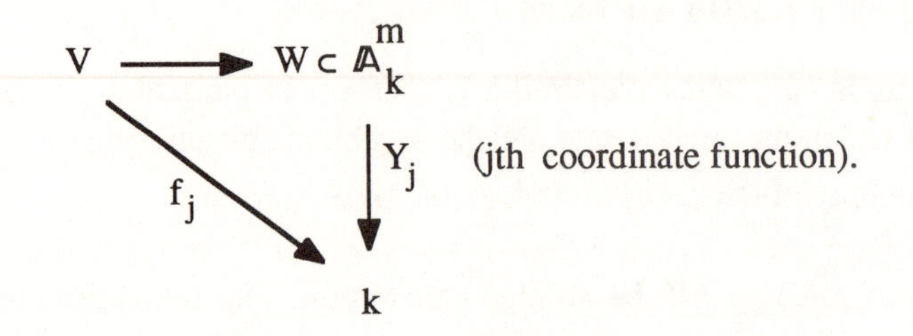

(jth coordinate function).

This is clear: if f is given by $F_1, .. F_m$, then the composite is just $P \mapsto F_j(P)$, which is a polynomial function. Conversely, if $f_j \in k[V]$ for each j, then for any choice of $F_j \in k[X_1, .. X_n]$ such that $f_j = F_j \mod I(V)$, I get a description of f as the polynomial map given by $(F_1, .. F_m)$.

In view of this claim, the map f can be written $f = (f_1, .. f_m)$.

The composite of polynomial functions is defined in the obvious way: if $V \subset \mathbb{A}^n$, $W \subset \mathbb{A}^m$ and $U \subset \mathbb{A}^\ell$ are algebraic sets, and $f\colon V \to W$ and $g\colon W \to U$ are polynomial maps, then $g \circ f\colon V \to U$ is again a polynomial map; for if f is given by $F_1, .. F_m \in k[X_1, .. X_n]$, and g is given by $G_1, .. G_\lambda \in k[Y_1, .. Y_m]$, then $g \circ f$ is given by

$$G_1(F_1, .. F_m), .. G_\ell(F_1, .. F_m) \in k[X_1, .. X_n].$$

**Definition.** A polynomial map $f\colon V \to W$ between algebraic sets is an *isomorphism* if there exists a polynomial map $f\colon W \to V$ such that $f \circ g = g \circ f = \mathrm{id}$.

Several examples of polynomial maps have already been given: the parametrisations $\mathbb{R}^1 \to C \subset \mathbb{R}^2$ by $t \mapsto (t^2, t^3)$ or $(t^2 - 1, t^3 - t)$ given in (2.1), and the map $k \to C \subset \mathbb{A}^3{}_k$ by $t \mapsto (t^3, t^4, t^5)$ discussed in (3.11, (b)) are clearly of this kind. Also, while discussing Noether normalisation, I had an algebraic set $V \subset \mathbb{A}^n{}_k$, and considered the general projection $p\colon V \to \mathbb{A}^m{}_k$ defined by m 'fairly general' linear forms $Y_1, .. Y_m$; since the $Y_i$ are linear forms in the coordinates $X_i$ of $\mathbb{A}^m{}_k$, this is a polynomial map.

On the other hand the parametrisation of the circle given in (1.1) is given by rational functions (there's a term $(\lambda^2 + 1)$ in the denominator); and the inverse map $(X, Y) \dashrightarrow T = Y/X$ from either of the singular cubics $C \subset \mathbb{R}^2$ back to $\mathbb{R}^1$ is also disqualified (or at least, doesn't qualify *as written* ) for the same reason.

**(4.4) Polynomial maps and** k[V].

**Theorem.** Let $V \subset \mathbb{A}^n{}_k$ and $W \subset \mathbb{A}^m{}_k$ be algebraic sets as above.

(I) A polynomial map $f: V \to W$ induces a ring homomorphism $f^*: k[W] \to k[V]$, defined by composition of functions; that is, if $g \in k[W]$ is a polynomial function then so is $f^*(g) = g \circ f$, and $g \mapsto g \circ f$ defines a ring homomorphism, in fact a k-algebra homomorphism $f^*: k[W] \to k[V]$. (Note that it goes backwards.)

(II) Conversely, any k-algebra homomorphism $\Phi: k[W] \to k[V]$ is of the form $\Phi = f^*$ for a uniquely defined polynomial map $f: V \to W$.

Thus (I) and (II) show that

$$\{\text{polynomial maps } f: V \to W\} \longrightarrow \{\text{k-algebra homs. } \Phi: k[W] \to k[V]\}$$

by

$$f \longmapsto f^*$$

is a bijection.

(III) If $f: V \to W$ and $g: W \to U$ are polynomial maps then the two ring homomorphisms $(g \circ f)^* = f^* \circ g^* : k[U] \to k[V]$ coincide.

**Proof.** (I) By what I said in (4.3), $f^*(g)$ is a polynomial map $V \to k$, hence $f^*(g) \in k[V]$. Obviously $f^*(a) = a$ for all $a \in k$ (since $k$ is being considered as the constant functions on $V$, $W$). Finally the fact that $f^*$ is a ring homomorphism is formal, since both $k[W]$ and $k[V]$ are rings of functions. (The ring structure is defined pointwise, so for example, for $g_1, g_2 \in k[W]$, the sum $g_1 + g_2$ is defined as the function on $W$ such that $(g_1 + g_2)(P) = g_1(P) + g_2(P)$ for all $P \in W$; therefore $f^*(g_1 + g_2)(Q) = (g_1 + g_2)(f(Q)) = g_1(f(Q)) + g_2(f(Q)) = f^*g_1(Q) + f^*g_2(Q)$. No-one's going to read this rubbish, are they?)

(III) is just the fact that composition of maps is associative.

(II) is a little more tricky to get right, although it's still content-free. For $i = 1,.. m$, let $y_i \in k[W]$ be the ith coordinate function on $W$, so that

$$k[W] = k[y_1,.. y_m] = k[Y_1,.. Y_m]/I(W).$$

Now $\Phi: k[W] \to k[V]$ is given, so I can define $f_i \in k[V]$ by $f_i = \Phi(y_i)$.

Consider the map $f: V \to \mathbb{A}^m{}_k$ defined by $f(P) = f_1(P),.. f_m(P)$. This is a polynomial map since $f_i \in k[V]$. Furthermore, I claim that $f$ takes $V$ into $W$, that is, $f(V) \subset W$. Indeed, suppose that $G \in I(W) \subset k[Y_1,.. Y_m]$; then

$$G(y_1,.. y_m) = 0 \in k[W],$$

where the left-hand side means that I substitute the ring elements $y_i$ into the polynomial expression $G$. Therefore, $\Phi^*(G(y_1,.. y_m)) = 0 \in k[W]$; but $\Phi$ is a k-algebra homomorphism, so that

$$k[V] \ni 0 = \Phi^*(G(y_1,.. y_m)) = G(\Phi^*(y_1),.. \Phi^*(y_m)) = G(f_1,.. f_m).$$

The $f_i$ are functions on $V$, and $G(f_1,.. f_m) \in k[V]$ is by definition the function $P \mapsto G(f_1(P),.. f_m(P))$. This proves that for $P \in V$, and for every $G \in I(W)$, the coordinates $(f_1(P),.. f_m(P))$ of $f(P)$ satisfy $G(f_1(P),.. f_m(P)) = 0$. Since $W$ is the subset of $\mathbb{A}^m{}_k$ defined by the vanishing of $G \in I(W)$, it follows that $f(P) \in W$. This proves that $f$ given above is a polynomial map $f: V \to W$. To check that the two $k$-algebra homomorphisms $f^*, \Phi: k[W] \to k[V]$ coincide, it's enough to check that they agree on the generators, that is $f^*(y_i) = \Phi(y_i)$; a minute inspection of the construction of $f$ (at the start of the proof of (II) above) will reveal that this is in fact the case. An exactly similar argument shows that the map $f$ is uniquely determined by the condition $f^*(y_i) = \Phi(y_i)$. Q.E.D.

**(4.5) Corollary.** A polynomial map $f: V \to W$ is an isomorphism if and only if $f^*: k[W] \to k[V]$ is an isomorphism.

**Example.** Over an infinite field $k$, the polynomial map

$$\varphi: \mathbb{A}^1{}_k \to C: (Y^2 = X^3) \subset \mathbb{A}^2{}_k \quad \text{by} \quad T \mapsto (T^2, T^3)$$

is not an isomorphism. For in this case, the homomorphism

$$\varphi^*: k[C] = k[X, Y]/(Y^2 - X^3) \to k[T]$$

is given by $X \mapsto T^2$, $Y \mapsto T^3$. The image of $\varphi^*$ is the $k$-algebra generated by $T^2, T^3$, that is $k[T^2, T^3] \subsetneq k[T]$. (Please make sure you understand why $T^2, T^3$ don't generate $k[T]$; I can't help you on this.)

Notice that $\varphi$ is bijective, and so has a perfectly good inverse map $\psi: C \to \mathbb{A}^1{}_k$ given by $(X, Y) \mapsto 0$ if $X = Y = 0$, $Y/X$ otherwise. So why isn't $\varphi$ an isomorphism? The point is that $C$ has fewer polynomial functions on it than $\mathbb{A}^1$; in a sense you can see that for yourself, since $k[\mathbb{A}^1] = k[T]$ has a polynomial function with non-zero derivative at $0$. The gut feeling is that $\varphi$ 'squashes up the tangent vector at $0$'.

**(4.6) Affine variety.** Let $k$ be an field; I want an *affine variety* to be an irreducible algebraic subset $V \subset \mathbb{A}^n{}_k$, defined up to isomorphism.

Proposition 4.4 tells us that the coordinate ring $k[V]$ is an invariant of the isomorphism class of $V$. This allows me to give a definition of a variety making less use of the ambient space $\mathbb{A}^n{}_k$; the reason for wanting to do this is rather obscure, and for practical purposes the reader will not miss much if he ignores it:

subsequent references to an affine variety will always be taken in the sense given above (GOTO (4.7)).

**Definition.** An affine variety over a field $k$ is a set $V$, together with a ring $k[V]$ of $k$-valued functions $f\colon V \to k$ such that

(i) $k[V]$ is a finitely generated $k$-algebra,

and

(ii) for some choice $x_1,.. x_n$ of generators of $k[V]$ over $k$, the map

$$V \to \mathbb{A}^n_k$$

by

$$P \mapsto x_1(P),.. x_n(P)$$

embeds $V$ as an irreducible algebraic set.

**(4.7) Function field.** Let $V$ be an affine variety; then the coordinate ring $k[V]$ of $V$ is an integral domain whose elements are $k$-valued functions of $V$.

**Definition.** The *function field* $k(V)$ of $V$ is the field of fractions $k(V) = \mathrm{Quot}(k[V])$ of $k[V]$. An element $f \in k(V)$ is a *rational function* on $V$; note that $f \in k(V)$ is by definition a quotient $f = g/h$ with $g, h \in k[V]$ and $h \neq 0$.

A priori $f$ is not a function on $V$, because of the zeros of $h$; however, $f$ is well-defined at $P \in V$ whenever $h(P) \neq 0$, so is at least a 'partially defined function'. I now introduce terminology to shore up this notion.

**Definition.** Let $f \in k(V)$ and $P \in V$; I say that $f$ is *regular* at $P$, or that $P$ is in the *domain of definition* of $f$ if there exists an expression $f = g/h$ with $g, h \in k[V]$ and $h(P) \neq 0$.

An important point to bear in mind is that usually $k[V]$ will not be a UFD, so that $f \in k(V)$ may well have essentially different representations as $f = g/h$; see Ex. 4.9 for an example.

Write

$$\mathrm{dom}\, f = \{P \in V \mid f \text{ is regular at } P\}$$

for the *domain of definition* of $f$, and

$$\mathcal{O}_{V,P} = \{f \in k(V) \mid f \text{ is regular at } P\} = k[V][\{h^{-1} \mid h(P) \neq 0\}].$$

Then $\mathcal{O}_{V,P} \subset k(V)$ is a subring, the *local ring* of $V$ at $P$.

**(4.8) Theorem.** (I) $\mathrm{dom}\, f$ is open and dense in the Zariski topology.
Suppose that the field $k$ is algebraically closed; then

(II) $\operatorname{dom} f = V \iff f \in k[V]$;

(that is *polynomial function* = *regular rational function*). Furthermore, for any $h \in k[V]$ let

$$V_h = V - V(h) = \{P \in V \mid h(P) \neq 0\};$$

then

(III) $\operatorname{dom} f \supset V_h \iff f \in k[V][h^{-1}]$.

**Proof.** Define the *ideal of denominators* of $f \in k(V)$ by

$$D_f = \{h \in k[V] \mid hf \in k[V]\} \subset k[V]$$

$$= \{h \in k[V] \mid \exists \text{ an expression } f = g/h \text{ with } g \in k[V]\} \cup \{0\}.$$

From the first line, $D_f$ is obviously an ideal of $k[V]$. Then formally,

$$V - \operatorname{dom} f = \{P \in V \mid h(P) = 0 \text{ for all } h \in D_f\} = V(D_f),$$

so that $V - \operatorname{dom} f$ is an algebraic set of $V$; hence $\operatorname{dom} f = V - V(D_f)$ is the complement of a closed set, so open in the Zariski topology. It is obvious that $\operatorname{dom} f$ is non-empty, hence dense by Proposition 4.2.

Now using (b) of the Nullstellensatz,

$$\operatorname{dom} f = V \iff V(D_f) = \emptyset \iff 1 \in D_f, \text{ that is, } f \in k[V].$$

Finally,

$$\operatorname{dom} f \supset V_h \iff h \text{ vanishes on } V(D_f),$$

and using (c) of the Nullstellensatz,

$$\iff h^n \in D_f \text{ for some } n, \text{ that is, } f = g/h^n \in k[V][h^{-1}]. \quad \text{Q.E.D.}$$

**(4.9) Rational maps.**

Let $V$ be an affine variety.

**Definition.** A *rational map* $f\colon V \dashrightarrow \mathbb{A}^n{}_k$ is a partially defined map given by rational functions $f_1,.. f_n$, that is,

$$f(P) = f_1(P),.. f_n(P) \quad \text{for all} \quad P \in \bigcap \operatorname{dom} f_i.$$

By definition, $\operatorname{dom} f = \bigcap \operatorname{dom} f_i$; as before, $F$ is said to be *regular* at $P \in V$ if and only if $P \in \operatorname{dom} f$. A rational map $V \dashrightarrow W$ between two affine varieties $V \subset \mathbb{A}^n$ and $W \subset \mathbb{A}^m$ is defined to be a rational map $f\colon V \dashrightarrow \mathbb{A}^n$ such that $f(\operatorname{dom} f) \subset W$.

Two examples of rational maps were described at the end of (4.3).

**(4.10) Composition of rational maps.** The composite $g \circ f$ of rational maps $f: V \dashrightarrow W$ and $g: W \dashrightarrow U$ may not be defined. This is a difficulty caused by the fact that a rational map is not a map: in a natural and obvious sense, the composite is a map defined on $\operatorname{dom} f \cap f^{-1}(\operatorname{dom} g)$; however, it can perfectly well happen that this is empty (see Ex. 4.10).

Expressed algebraically, the same problem also occurs: suppose that $f$ is given by $f_1,.. f_m \in k(V)$, so that

$$f: V \dashrightarrow W \subset \mathbb{A}^m$$

by

$$P \mapsto f_1(P),.. f_m(P)$$

for $P \in \bigcap \operatorname{dom} f_i$; any $g \in k[W]$ is of the form $g = G \mod I(W)$ for some $G \in k[Y_1,.. Y_m]$, and $g \circ f = G(f_1,.. f_m)$ is well-defined in $k(V)$. So exactly as in (4.4), there is a k-algebra homomorphism

$$f^*: k[W] \to k(V)$$

corresponding to $f$. However, if $h \in k[W]$ is in the kernel of $f^*$, then no meaning can be attached to $f^*(g/h)$, so that $f^*$ cannot be extended to a field homomorphism $k(W) \to k(V)$.

**Definition.** $f: V \dashrightarrow W$ is *dominant* if $f(\operatorname{dom} f)$ is dense in $W$ for the Zariski topology.

Geometrically, this means that $f^{-1}(\operatorname{dom} g) \subset \operatorname{dom} f$ is a dense open set for any rational map $g: W \dashrightarrow U$, so that $g \circ f$ is defined on a dense open set of $V$, so is a partially defined map $V \dashrightarrow U$.

Algebraically,

$$f \text{ is dominant} \iff f^*: k[V] \to k(W) \text{ is injective.}$$

For given $g \in k[W]$,

$$g \in \ker f^* \iff f(\operatorname{dom} f) \subset V(g),$$

that is, $f^*$ is not injective if and only if $f(\operatorname{dom} f)$ is contained in a strict algebraic subset of $W$.

Clearly, the composite $g \circ f$ of rational maps $f$ and $g$ is defined provided that $f$ is dominant: $g \circ f$ is the rational map whose components are $f^*(g_i)$. Notice that the domain of $g \circ f$ certainly contains $f^{-1}(\operatorname{dom} g) \cap \operatorname{dom} f$, but may very well be larger (see Ex. 4.6).

**(4.11) Theorem.** (I) A dominant rational map $f\colon V \dashrightarrow W$ defines a field homomorphism $f^*\colon k(W) \to k(V)$.

(II) Conversely, a $k$-homomorphism $\Phi\colon k(W) \to k(V)$ comes from a uniquely defined dominant rational map $f\colon V \dashrightarrow W$.

(III) If $f$ and $g$ are dominant then $(g \circ f)^* = f^* \circ g^*$.

The proof requires only minor modifications to that of (4.4).

**(4.12) Morphisms from an open subset of an affine variety.** Let $V$, $W$ be affine varieties, and $U \subset V$ an open subset.

**Definition.** A *morphism* $f\colon U \to W$ is a rational map $f\colon V \dashrightarrow W$ such that $U \subset \operatorname{dom} f$, so that $f$ is regular at every $P \in U$.

If $U_1 \subset V$ and $U_2 \subset W$ are opens, then a morphism $f\colon U_1 \to U_2$ is just a morphism $f\colon U_1 \to W$ such that $f(U_1) \subset U_2$. An isomorphism is a morphism which has a two-sided inverse morphism.

Note that if $V$, $W$ are affine varieties, then by Theorem 4.8, (II),

$$\{\text{morphisms } f\colon V \to W\} = \{\text{polynomial maps } f\colon V \to W\};$$

the left-hand side of the equation consists of rational objects subject to regularity conditions, whereas the right-hand side is more directly in terms of polynomials.

**Example.** The parametrisation of the cuspidal cubic $\mathbb{A}^1 \to C\colon (Y^2 = X^3)$ of (2.1) induces an isomorphism $\mathbb{A}^1 \setminus \{0\} \cong C \setminus \{(0,0)\}$; see Ex. 4.5 for details.

**(4.13) Standard open subsets.** Let $V$ be an affine variety. For $f \in k[V]$, write $V_f$ for the open set $V_f = V \setminus V(f) = \{P \in V \mid f(P) \neq 0\}$. The $V_f$ are called *standard open sets* of $V$.

**Proposition.** $V_f$ is isomorphic to an affine variety, and

$$k[V_f] = k[V][f^{-1}].$$

**Proof.** The idea is to consider the graph of the function $f^{-1}$; a similar trick was used for (b) $\Rightarrow$ (c) in the proof of NSS (3.10).

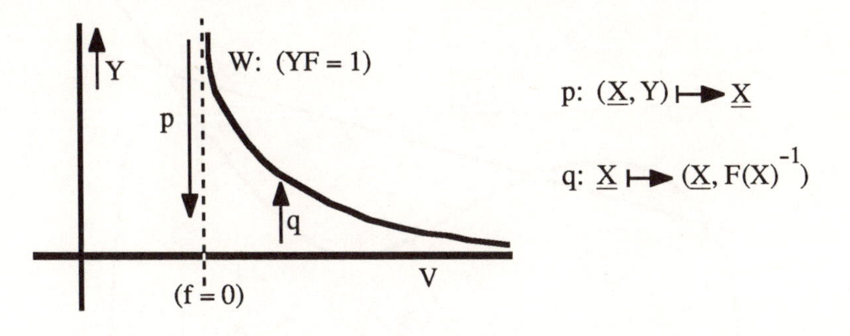


Let $J = I(V) \subset k[X_1,.. X_n]$, and choose $F \in k[X_1,.. X_n]$ such that $f = F$ mod $I(V)$. Now define $I = (J, YF - 1) \subset k[X_1,.. X_n, Y]$, and let

$$V(I) = W \subset \mathbb{A}^{n+1}.$$

It is easy to check that the maps indicated in the diagram are inverse morphisms between $W$ and $V_f$. The statement about the coordinate ring is contained in (4.8, (III)). Q.E.D.

The standard open sets $V_f$ are important because they form a basis for the Zariski topology of $V$: every open set $U \subset V$ is a union of $V_f$'s (since every closed subset is of the form $V(I) = \bigcap_{f \in I} V(f)$ for some ideal). Thus the point of the result just proved is that every open set $U \subset V$ is a union of open sets $V_f$ which are affine varieties.

**(4.14) Worked example.** In §2 I discussed the addition law $(A, B) \mapsto A + B$ on a plane non-singular (projective) cubic $C \subset \mathbb{P}^2$. Let $C_0$: $(y^2 = x^3 + ax + b)$ be a non-singular affine cubic:

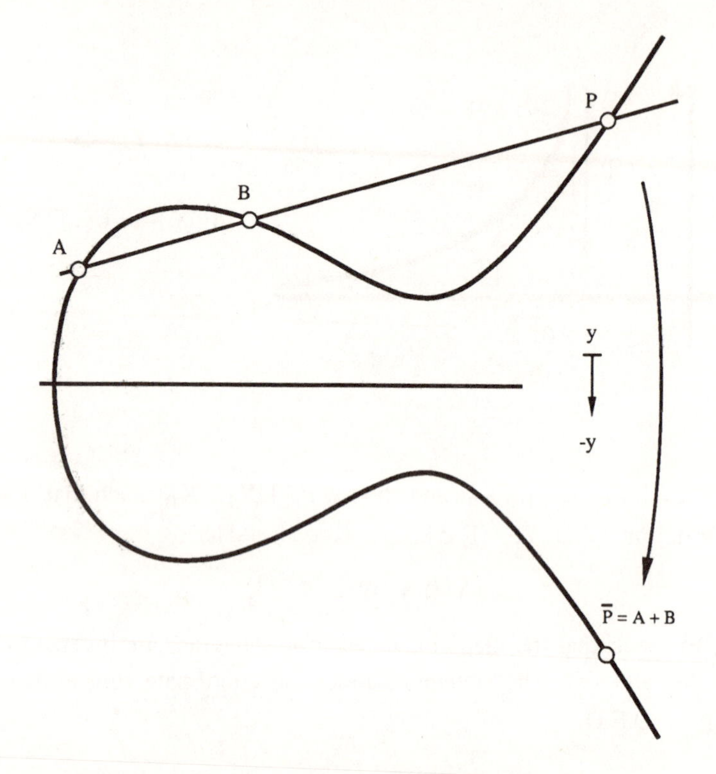


I show here that the addition law defines a rational map $\varphi: C_0 \times C_0 \dashrightarrow C_0$, and that $\varphi$ is a morphism wherever it should be. Although I will not labour the point, this argument can be used to give another proof 'by continuity' of the associativity of the group law valid for any field (see the discussion in (2.10)).

It is not difficult to see (compare Ex. 2.7) that if $A = (x, y)$, $B = (x', y')$, and $x \neq x'$ then setting $u = (y - y')/(x - x')$, the 3rd point of intersection is $P = (x'', y'')$, where

$$x'' = f(x, y, x', y') = u^2 - (x + x'),$$

$$y'' = g(x, y, x', y') = u^3 + xu + y'.$$

Since $x''$ and $y''$ are rational functions in the coordinates $(x, y)$, $(x', y')$, this shows that $\varphi: C_0 \times C_0 \dashrightarrow C_0 \times C_0$ is a rational map. From the given formula, $\varphi$ is a morphism wherever $x \neq x'$, since then the denominator of $u$ is non-zero. Now if $x = x'$ and $y = -y'$, then $x''$ and $y''$ should be infinity, corresponding to the fact that the line AB meets the projective curve C at the point at infinity $O = (0,1,0)$.

However, if $x = x'$ and $y = y' \neq 0$ then the point $P = (x'', y'')$ should be well-defined. I claim that f, g are regular functions on $C_0 \times C_0$ at such points: to see this, note that

$$y^2 = x^3 + ax + b \quad\text{and}\quad y'^2 = x'^3 + ax' + b,$$

giving

$$y^2 - y'^2 = x^3 - x'^3 + a(x - x');$$

therefore as rational functions on $C_0 \times C_0$, there is an equality

$$u = (y - y')/(x - x') = (x^2 + xx' + x'^2 + a)/(y + y').$$

Looking at the denominator, it follows that u (hence also f and g) is regular whenever $y \neq -y'$.

The conclusion of the calculation is the following proposition: the addition law $\varphi\colon C_0 \times C_0 \dashrightarrow C_0$ is a morphism at $(A, B) \in C_0 \times C_0$ provided that $A + B \neq O$.

## Exercises to §4.

**4.1.** Check that the statements of §4 up to and including (4.8, (I)) are valid for any field k; discover in particular what they mean for a finite field. Give a counter-example to (4.8, (II)) if k is not algebraically closed.

**4.2.** $\varphi\colon \mathbb{A}^1 \to \mathbb{A}^3$ is the polynomial map given by $X \mapsto (X, X^2, X^3)$; prove that the image of $\varphi$ is an algebraic subset $C \subset \mathbb{A}^3$ and that $\varphi\colon \mathbb{A}^1 \to C$ is an isomorphism. Try to generalise.

**4.3.** $\varphi_n\colon \mathbb{A}^1 \to \mathbb{A}^2$ is the polynomial map given by $X \mapsto (X^2, X^n)$; show that if n is even, the image of $\varphi_n$ is isomorphic to $\mathbb{A}^1$, and $\varphi_n$ is 2-to-1 outside 0. And if n is odd, show that $\varphi_n$ is bijective, and give a rational inverse of $\varphi_n$.

**4.4.** Prove that a morphism $\varphi\colon X \to Y$ between two affine varieties is an isomorphism of X with a subvariety $\varphi(X) \subset Y$ if and only if the induced map $\Phi\colon k[Y] \to k[X]$ is surjective.

**4.5.** Let $C\colon (Y^2 = X^3) \subset \mathbb{A}^2$; then

(a) the parametrisation $f\colon \mathbb{A}^1 \to C$ given by $(T^2, T^3)$ is a polynomial map;

(b) f has a rational inverse $g\colon C \dashrightarrow \mathbb{A}^1$ defined by $(X, Y) \mapsto Y/X$;

(c) $\operatorname{dom} g = C \setminus \{(0,0)\}$;

(d) f and g give inverse isomorphisms $\mathbb{A}^1 \setminus \{0\} \cong C \setminus \{(0,0)\}$.

**4.6.** (i) Show that the domain of $g \circ f$ may be strictly larger than $\operatorname{dom} f \cap f^{-1}(\operatorname{dom} g)$. (Hint: this may happen if g and f are inverse rational maps; try f and g as in Ex. 4.5.)

(ii) Most courses on calculus of several variables contain examples such as the function $f(x, y) = xy/(x^2 + y^2)$. Explain how come $f$ is $C^\infty$ when restricted to any smooth curve through $(0, 0)$, but is not even continuous as a function of 2 variables.

**4.7.** Let $C: (Y^2 = X^3 + X^2) \subset \mathbb{A}^2$; the familiar parametrisation $\varphi: \mathbb{A}^1 \to C$ given by $(T^2 - 1, T^3 - T)$ is a polynomial map, but is not an isomorphism (why not?). Find out whether the restriction $\varphi': \mathbb{A}^1 \setminus \{1\} \to C$ is an isomorphism:

**4.8.** Let $C: (Y^3 = X^4 + X^3) \subset \mathbb{A}^2$; show that $(X, Y) \mapsto X/Y$ defines a rational map $\psi: C \dashrightarrow \mathbb{A}^1$, and that its inverse is a polynomial map $\varphi: \mathbb{A}^1 \to C$ parametrising $C$. Prove that $\varphi$ restricts to an isomorphism

$$\mathbb{A}^1 \setminus \{3 \text{ pts.}\} \cong C \setminus \{(0,0)\}.$$

**4.9.** Let $V: (XT = YZ) \subset \mathbb{A}^4$; explain why $k[V]$ is not a UFD. (It's not hard to get the idea, but rather harder to give a rigorous proof). If $f = X/Y \in k(V)$, find $\operatorname{dom} f$, and prove that it is strictly bigger than the locus $(Y = 0) \subset V$.

**4.10.** Let $f: \mathbb{A}^1 \to \mathbb{A}^2$ be given by $X \mapsto (X, 0)$, and let $g: \mathbb{A}^2 \dashrightarrow \mathbb{A}^1$ be the rational map given by $(X, Y) \mapsto X/Y$; show that the composite $g \circ f$ is not defined anywhere. Determine what is the largest subset of the function field $k(\mathbb{A}^1)$ on which $g^*$ is defined.

**4.11.** Define and study the notion of product of two algebraic sets. More precisely,

(i) if $V \subset \mathbb{A}^n_k$ and $W \subset \mathbb{A}^m_k$ are algebraic sets, prove that $V \times W \subset \mathbb{A}^{n+m}_k$ is also;

(ii) give examples to show that the Zariski topology on $V \times W$ is not the product topology of those on $V$ and on $W$;

(iii) prove that $V, W$ irreducible $\implies V \times W$ irreducible;

(iv) prove that if $V \cong V'$ and $W \cong W'$ then $V \times W \cong V' \times W'$.

**4.12.** (a) Prove that any $f \in k(\mathbb{A}^2)$ which is not regular at the origin $(0, 0)$ also fails to be regular at points of a curve passing through $(0, 0)$.

(b) Deduce that $\mathbb{A}^2 \setminus (0, 0)$ is not affine.

(Hints: For (a), use the fact that $k(\mathbb{A}^2) = k[X, Y]$ is the field of fractions of the UFD $k[X, Y]$, together with the result of Ex. 3.13, (b). For (b), assume that $\mathbb{A}^2 \setminus (0, 0)$ is affine, and determine its coordinate ring; then get a contradiction using Corollary 4.5.)

# Chapter III. Applications

## §5. Projective and birational geometry

The first part of §5 aims to generalise the content of §§3–4 to projective varieties; this is fairly mechanical, with just a few essential points. The remainder of the section is concerned with birational geometry, taking up the function field $k(V)$ from the end of §4; this is material which fits equally well into the projective or affine context.

**(5.0) Why projective varieties?** The cubic curve

$$C\colon (Y^2Z = X^3 + aXZ^2 + bZ^3) \subset \mathbb{P}^2$$

is the union of two affine curves

$$C_0\colon (y^2 = x^3 + ax + b) \subset \mathbb{A}^2 \quad \text{(the piece } (Z = 1) \text{ of } C)$$

and

$$C_1\colon (z_1 = x_1^3 + axz_1^2 + bz_1^3) \subset \mathbb{A}^2 \quad \text{(the piece } (Y = 1)),$$

glued together by the isomorphism

$$C_0 \setminus (y = 0) \longrightarrow C_1 \setminus (z_1 = 0)$$

by

$$(x, y) \mapsto (x/y, 1/y).$$

As a much simpler example, $\mathbb{P}^1$ with homogeneous coordinates $(X, Y)$ is the union of 2 copies of $\mathbb{A}^1$ with coordinates $x_0, y_1$ respectively, glued together by the isomorphism

$$\mathbb{A}^1 \setminus (x_0 = 0) \to \mathbb{A}^1 \setminus (y_1 = 0)$$

by

$$x_0 \mapsto 1/x_0.$$

The usual picture is

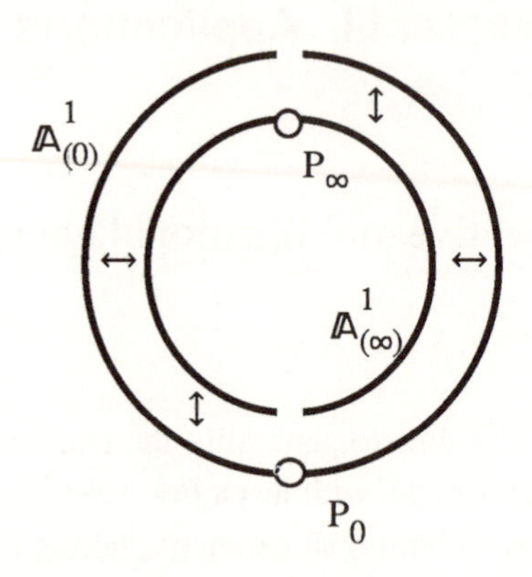


(the arrows $\longleftrightarrow$ denote glueing).

It's important to understand that *these varieties are strictly bigger than any affine variety*. In fact, with the natural notion of morphism (to be introduced shortly), it can be seen that there are no non-constant morphisms $\mathbb{P}^1 \to \mathbb{A}^n$ or $C \to \mathbb{A}^n$ for any $n$ (see Ex. 5.1 and Ex. 5.12, and the discussion in (8.10)).

One solution to this problem is to define the notion of 'abstract variety' $V$ as a union $V = \bigcup V_i$ of affine varieties, modulo suitable glueing. By analogy with the definition of manifolds in topology, this is an attractive idea, but it leads to many more technical difficulties. Using projective varieties sidesteps these problems by working in the ready-made ambient space $\mathbb{P}^n$, so that (apart from a little messing about with homogeneous polynomials) they are not much harder to study than affine varieties. In fact, although this may not be clear at an elementary level, projective varieties to a quite remarkable extent provide a natural framework for studying varieties (this is briefly discussed from a more high-brow point of view in (8.11)).

**(5.1) Graded rings and homogeneous ideals.**

**Definition.** A polynomial $f \in k[X_0,.. X_n]$ is *homogeneous of degree* $d$ if

$$f = \sum a_{i_0 .. i_n} X_0^{i_0} .. X_n^{i_n} \quad \text{with} \quad a_{i_0 .. i_n} \neq 0 \text{ only if } i_0 + .. i_n = d.$$

Any $f \in k[X_0,.. X_n]$ has a unique expression $f = f_0 + f_1 + .. f_N$ in which $f_d$ is homogeneous of degree $d$ for each $d = 0, 1,.. N$.

**Proposition.** If $f$ is homogeneous of degree $d$ then

$$f(\lambda X_0,.. \lambda X_n) = \lambda^d f(X_0,.. X_n) \quad \text{for all} \quad \lambda \in k;$$

if $k$ is an infinite field then the converse also holds.

**Proof.** Try it and see.

**Definition.** An ideal $I \subset k[X_0,.. X_n]$ is *homogeneous* if for all $f \in I$, the homogeneous decomposition $f = f_0 + f_1 + .. f_N$ of $f$ satisfies $f_i \in I$ for all i.

It is equivalent to say that $I$ is generated by (finitely many) homogeneous polynomials.

**(5.2) The homogeneous V-I correspondences.**

Let $\mathbb{P}^n{}_k$ be the n-dimensional projective space over a field $k$, with $X_0,.. X_n$ as homogeneous coordinates. Then $f \in k[X_0,.. X_n]$ is *not* a function on $\mathbb{P}^n{}_k$: by definition, $\mathbb{P}^n{}_k = k^{n+1} \setminus \{0\}/\sim$, where $\sim$ is the equivalence relation given by $(X_0,.. X_n) \sim (\lambda X_0,.. \lambda X_n)$ for $\lambda \in k \setminus \{0\}$; $f$ is a function on $k^{n+1}$. Nevertheless, for $P \in \mathbb{P}^n$, the condition $f(P) = 0$ is well-defined provided that $f$ is homogeneous: suppose $P = (X_0: .. X_n)$, so that $(X_0,.. X_n)$ is a representative in $k^{n+1} \setminus \{0\}$ of the equivalence class of $P$. Then since $f(\lambda \underline{X}) = \lambda^d f(\underline{X})$, if $f(X_0,.. X_n) = 0$ then also $f(\lambda X_0,.. \lambda X_n) = 0$, so that the condition $f(P) = 0$ is independent of the choice of representative. With this is mind, define as before correspondences

$$\{\text{homog. ideals } J \subset k[X_0,.. X_n]\} \xleftrightarrow{V, I} \{\text{subsets } X \subset \mathbb{P}^n{}_k\}$$

by

$$V(J) = \{P \in \mathbb{P}^n{}_k \mid f(P) = 0 \ \forall \text{ homogeneous elements } f \in J\}$$

and

$$I(X) = \{f \in k[X_0,.. X_n] \mid f(P) = 0 \text{ for all } P \in X\}.$$

As an exercise, check that you understand why $I(X)$ is a homogeneous ideal.

The correspondences $V$ and $I$ satisfy the same formal properties as the affine $V$ and $I$ correspondences introduced in §3 (for example $V(J_1 + J_2) = V(J_1) \cap V(J_2)$). A subset of the form $V(I)$ is an *algebraic subset* of $\mathbb{P}^n{}_k$, and as in the affine case, $\mathbb{P}^n{}_k$ has a *Zariski topology* in which the closed sets are the algebraic subsets.

**(5.3) Projective Nullstellensatz.** As with the affine correspondences, it is purely formal that $I(V(J)) \supset \operatorname{rad} J$ for any ideal $J$, and that for an algebraic set, $V(I(X)) = X$. There's just one point where care is needed: the trivial ideal $(1) = k[X_0,.. X_n]$ (the whole ring) defines the empty set in $k^{n+1}$, hence also in $\mathbb{P}^n{}_k$, which is as it should be; however, the ideal $(X_0,.. X_n)$ defines $\{0\}$ in $k^{n+1}$, which also corresponds to the empty set in $\mathbb{P}^n_k$. The ideal $(X_0,.. X_n)$ is an awkward (empty-set theoretical) exception to several statements in the theory, and is traditionally known as the 'irrelevant ideal'.

The homogeneous version of the Nullstellensatz thus becomes

**Theorem.** Assume that $k$ is an algebraically closed field. Then

(i) $V(J) = \emptyset \iff \text{rad } J \supset (X_0,.. X_n)$;

(ii) if $V(J) \neq \emptyset$ then $I(V(J)) = \text{rad } J$.

**Corollary.** I and V determine inverse bijections

$$\left\{\begin{matrix}\text{homogeneous radical}\\ \text{ideals } J \subset k[x_0,.. x_n]\\ \text{with } J \neq k[x_0,.. x_n]\end{matrix}\right\} \longleftrightarrow \left\{\begin{matrix}\text{algebraic subsets}\\ X \subset \mathbb{P}^n\end{matrix}\right\}$$

$$\cup \qquad\qquad\qquad\qquad \cup$$

$$\left\{\begin{matrix}\text{homogeneous prime}\\ \text{ideals } J \subset k[x_0,.. x_n]\\ \text{with } J \neq k[x_0,.. x_n]\end{matrix}\right\} \longleftrightarrow \left\{\begin{matrix}\text{irreducible algebraic}\\ \text{subsets } X \subset \mathbb{P}^n\end{matrix}\right\}$$

**Proof.** Let $\pi\colon \mathbb{A}^{n+1} \setminus \{0\} \to \mathbb{P}^n$ be the map defining $\mathbb{P}^n$. For a homogeneous ideal $J \subset k[X_0,.. X_n]$, write (in temporary notation) $V^a(J) \subset \mathbb{A}^{n+1}$ for the affine algebraic set defined by $J$. Then since $J$ is homogeneous, $V^a(J)$ has the property

$$\alpha_0,.. \alpha_n \in V^a(J) \iff \lambda\alpha_0,.. \lambda\alpha_n \in V^a(J),$$

and $V(J) = V^a(J) \setminus \{0\}/\sim \ \subset \mathbb{P}^n$. Hence

$$V(J) = \emptyset \iff V^a(J) \subset \{0\} \iff \text{rad } J \supset (X_0,.. X_n),$$

where the last implication uses the affine Nullstellensatz. Also, if $V(J) \neq \emptyset$ then

$$f \in I(V(J)) \iff f \in I(V^a(J)) \iff f \in \text{rad } J. \qquad \text{Q.E.D.}$$

The affine subset $V^a(J)$ occuring above is called the *affine cone* over the projective algebraic subset $V(J)$.

**(5.4) Rational functions on V.** Let $V \subset \mathbb{P}^n_k$ be an irreducible algebraic set, and $I(V) \subset k[X_0,.. X_n]$ its ideal; there is no direct way of defining regular functions on $V$ in terms of polynomials: an element $F \in k[X_0,.. X_n]$ gives a function on the affine cone over $V$, but (by case $d = 0$ of Proposition 5.1) this will be constant on equivalence classes only if $F$ is homogeneous of degree $0$, that is, a constant. So from the start, I work with rational functions only:

**Definition.** A *rational function* on $V$ is a (partially defined) function $f\colon V \dashrightarrow k$ given by $f(P) = g(P)/h(P)$, where $g, h \in k[X_0,.. X_n]$ are homogeneous polynomials of the same degree $d$.

Note here that provided $h(P) \neq 0$, the quotient $g(P)/h(P)$ is well-defined, since

$$g(\lambda\underline{X})/h(\lambda\underline{X}) = \lambda^d g(\underline{X})/\lambda^d h(\underline{X}) = g(\underline{X})/h(\underline{X}) \quad \text{for } 0 \neq \lambda \in k.$$

Now obviously $g/h$ and $g'/h'$ define the same rational function on $V$ if and only if $h'g - g'h \in I(V)$, so that the set of all rational functions is the field

$$k(V) = \{g/h \mid g, h \in k[X_0,.. X_n] \text{ homog. of same degree, } h \notin I(V)\}/\sim,$$

where $\sim$ is the equivalence relation

$$g/h \sim g'/h' \iff h'g - g'h \in I(V).$$

$k(V)$ is the (rational) *function field* of $V$.

The following definitions are just as in the affine case. For $f \in k(V)$ and $P \in V$, say that $f$ is *regular* at $P$ if there exists an expression $f = g/h$, with $g, h$ homogeneous polynomials of the same degree, such that $h(P) \neq 0$. Write

$$\operatorname{dom} f = \{P \in V \mid f \text{ is regular at } P\}$$

and

$$\mathcal{O}_{V,P} = \{f \in k(V) \mid f \text{ is regular at } P\}.$$

Clearly, $\operatorname{dom} f \subset V$ is a dense Zariski open set in $V$ (the proof is as in (4.8, (I)), and $\mathcal{O}_{V,P} \subset k(V)$ is a subring.

**(5.5) Affine covering of a projective variety.** Let $V \subset \mathbb{P}^n$ be an irreducible algebraic set, and suppose for simplicity that $V \not\subset (X_i = 0)$ for any $i$. We know that $\mathbb{P}^n$ is covered by $(n + 1)$ affine pieces $\mathbb{A}^n_{(i)}$, with affine (inhomogeneous) coordinates $X_0^{(i)},.. X_{i-1}^{(i)}, X_{i+1}^{(i)},.. X_n^{(i)}$, where

$$X_j^{(i)} = X_j/X_i \quad \text{for } j \neq i.$$

Write $V_{(i)} = V \cap \mathbb{A}^n_{(i)}$. Then $V_{(i)} \subset \mathbb{A}^n_{(i)}$ is clearly an affine algebraic set, because

$$V_{(0)} \ni P = (1, x_1^{(0)},.. x_n^{(0)})$$

$$\iff \quad f(1, x_1^{(0)},.. x_n^{(0)}) = 0 \ \forall \text{ homogeneous } f \in I(V),$$

which is a set of polynomial relations in the coordinates $(x_1^{(0)},.. x_n^{(0)})$ of P. For clarity, I have taken $i = 0$ in the argument, and will continue to do so whenever convenient. The reader should remember that the same result applies to any of the other affine pieces $V_{(i)}$. The $V_{(i)}$ are called *standard affine pieces* of V.

**Proposition.** (i) The correspondence $V \mapsto V_{(0)} = V \cap \mathbb{A}^n_{(0)}$ gives a bijection

$$\left\{ \begin{matrix} \text{irreducible algebraic} \\ \text{subsets } V \subset \mathbb{P}^n \end{matrix} \middle| V \not\subset (X_0 = 0) \right\} \leftrightarrow \left\{ \begin{matrix} \text{irreducible algebraic} \\ \text{subsets } V_{(0)} \subset \mathbb{A}^n_{(0)} \end{matrix} \right\};$$

the inverse correspondence is given by taking closure in the Zariski topology.

(ii) Write $I^h(V) \subset k[X_0,.. X_n]$ for the homogeneous ideal of $V \subset \mathbb{P}^n$ introduced in this section and $I^a(V_{(0)}) \subset k[X_1,.. X_n]$ for the usual (as in §3) inhomogeneous ideal of $V_{(0)} \subset \mathbb{A}^n_{(0)}$; then $I^h(V)$ and $I^a(V_{(0)})$ are related as follows:

$$I^a = \{f(1, X_1,.. X_n) \mid f \in I^h(V)\},$$

and

$$I^h(V)_d = \{X_0^d f(X_1/X_0,.. X_n/X_0) \mid f \in I^a(V_{(0)}), \text{ with } \deg f \le d\},$$

where the subscript in $I^h(V)_d$ denotes the piece of degree d.

(iii) $k(V) \cong k(V_{(0)})$, and for $f \in k(V)$, the domain of f as a function on $V_{(0)}$ is $V_{(0)} \cap \operatorname{dom} f$.

**Proof.** (i) and (ii) are easy; (iii) If $f, g \in k[X_0,.. X_n]$ are homogeneous of degree d, and $g \notin I(V)$, then $f/g \in k(V)$ restricted to $V_{(0)}$ is the function

$$f(1, X_1/X_0,.. X_n/X_0)/g(1, X_1/X_0,.. X_n/X_0);$$

this defines a map $k(V) \to k(V_{(0)})$, and it's easy to see what its inverse is.

**(5.6) Rational maps and morphisms.** Rational maps between projective (or affine) varieties are defined using k(V): if $V \subset \mathbb{P}^n$ is an irreducible algebraic set, a rational map $V \dashrightarrow \mathbb{A}^m$ is a (partially defined) map given by $P \mapsto (f_1(P),.. f_m(P))$, where $f_1,.. f_m \in k(V)$. A rational map $V \dashrightarrow \mathbb{P}^m$ is defined by $P \mapsto (f_0(P) : f_1(P) : .. f_m(P))$ where $f_0, f_1,.. f_m \in k(V)$. Notice that if $0 \neq g \in k(V)$, then $gf_0, gf_1,.. gf_m$

defines the same rational map. Therefore, (assuming that $V$ does not map into the smaller projective space $(X_0 = 0)$), it would be possible to assume throughout that $f_0 = 1$.

Clearly then, there is a bijection between the two sets

$$\{\text{rational maps } f\colon V \dashrightarrow \mathbb{A}^m \subset \mathbb{P}^m\}$$

and

$$\{\text{rational maps } f\colon V \dashrightarrow \mathbb{P}^m \mid f(V) \not\subset (X_0 = 0)\},$$

since either kind of maps is given by $m$ elements $f_i \in k(V)$.

**Definition.** A rational map $f\colon V \dashrightarrow \mathbb{P}^m$ is *regular* at $P \in V$ if there exists an expression $f = (f_0, f_1, .. f_m)$ such that

(i) each of $f_0, .. f_m$ is regular at $P$;

and

(ii) at least one $f_i(P) \neq 0$.

The second condition is required here in order that the ratio between the $f_i$ is defined at $P$. If $f$ is regular at $P$ (as before, this is also expressed $P \in \operatorname{dom} f$) then $f\colon U \to \mathbb{A}^m_{(i)} \subset \mathbb{P}^m$ is a morphism for a suitable open neighbourhood $P \in U \subset V$: just take $U = \bigcap_j \operatorname{dom}(f_j/f_i)$ where $f_i(P) \neq 0$; then $f$ is the morphism given by $\{f_j/f_i\}_{j=0,1,..m}$.

If $U \subset V$ is an open subset of a projective variety $V$ then a *morphism* $f\colon U \to W$ is a rational map $f\colon V \dashrightarrow W$ such that $\operatorname{dom} f \supset U$. So a morphism is just a rational map that is everywhere regular on $U$.

**(5.7) Examples. (I) Rational normal curve.** This is a very easy example of an isomorphic embedding $f\colon \mathbb{P}^1 \xrightarrow{\cong} C \subset \mathbb{P}^m$ which generalises the parametrised conic of (1.7), and which occurs throughout projective and algebraic geometry. Define

$$f\colon \mathbb{P}^1 \to \mathbb{P}^m \quad \text{by} \quad (U : V) \mapsto (U^m : U^{m-1}V : .. V^m)$$

(writing down all monomials of degree $m$ in $U, V$). Arguing step-by-step:

(i) f is a rational map, since it's given by $((U/V)^m, (U/V)^{m-1}, .. 1)$;

(ii) $f$ is a morphism wherever $V \neq 0$ by the formula just written, and if $V = 0$ then $U \neq 0$, so a similar trick with $(V/U)$ works;

(iii) the image of $f$ is the set of points $(X_0 : .. X_m) \in \mathbb{P}^m$ such that

$$(X_0 : X_1) = (X_1 : X_2) = .. (X_{m-1} : X_m),$$

that is,

$$X_0X_2 = X_1^2, \ X_0X_3 = X_1X_2, \ X_0X_4 = X_1X_3,$$

etc. The equations can be written all together in the extremely convenient determinantal form

$$\text{rk} \begin{bmatrix} X_0 & X_1 & X_2 & .. & X_{m-1} \\ X_1 & X_2 & X_3 & .. & X_m \end{bmatrix} \le 1$$

(the rank condition means exactly that all $2 \times 2$ minors vanish). These are homogeneous equations defining an algebraic set $C \subset \mathbb{P}^m$;

(iv) the inverse morphism $g: C \to \mathbb{P}^1$ is not hard to find: just take a point of $C$ into the common ratio $(X_0 : X_1) = .. (X_{m-1} : X_m) \in \mathbb{P}^1$. As an exercise, find out for yourself what has to be checked, then check it all.

**(II) Linear projection, parametrising a quadric.** The map $\pi: \mathbb{P}^3 \dashrightarrow \mathbb{P}^2$ given by $(X_0, X_1, X_2, X_3) \mapsto (X_1, X_2, X_3)$ is a rational map, and a morphism outside the point $P_0 = (1, 0, 0, 0)$. Let $Q \subset \mathbb{P}^3$ be a quadric hypersurface with $P \in Q$. Then every point $P$ of $\mathbb{P}^2$ corresponds to a line $L$ of $\mathbb{P}^3$ through $P$, and $L$ should in general meet $Q$ at $P_0$ and a second point $\varphi(P)$: for example, if $Q: (X_0X_3 = X_1X_2)$, then $\pi_{|Q}: Q \dashrightarrow \mathbb{P}^2$ has the inverse map

$$\varphi: \mathbb{P}^2 \dashrightarrow Q \text{ given by } (X_1, X_2, X_3) \mapsto (X_1X_2/X_3, X_1, X_2, X_3).$$

This is essentially the same idea as the parametrisation of the circle in (1.1).

It is a rewarding exercise (see Ex. 5.2) to find $\text{dom}\,\pi$ and $\text{dom}\,\varphi$, and to give a geometrical interpretation of the singularities of $\pi$ and $\varphi$.

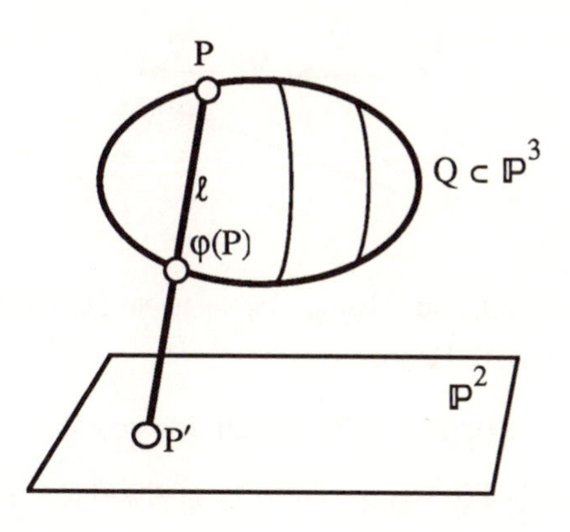


**(5.8) Birational maps.**
**Definition.** Let $V$ and $W$ be (affine or projective) varieties; then a rational map $f\colon V \dashrightarrow W$ is *birational* (or is a *birational equivalence*) if it has a rational inverse, that is, if there exists a rational map $g\colon W \dashrightarrow V$ such that $f \circ g = \mathrm{id}_W$ and $g \circ f = \mathrm{id}_V$.

**Proposition.** The following 3 conditions on a rational map $f\colon V \dashrightarrow W$ are equivalent:

(i) $f$ is a birational equivalence;

(ii) $f$ is dominant (see (4.10)), and $f^*\colon k(W) \to k(V)$ is an isomorphism;

(iii) there exist open sets $V_0 \subset V$ and $W_0 \subset W$ such that $f$ restricted to $V_0$ is an isomorphism $f\colon V_0 \to W_0$.

**Proof.** $f^*$ is defined in the same way as for affine varieties, and (i) $\iff$ (ii) is the same as (4.9). (iii) $\Rightarrow$ (i) is clear, since an isomorphism $f\colon V_0 \to W_0$ and its inverse $g = f^{-1}\colon W_0 \to V_0$ are by definition rational maps between $V$ and $W$.

The essential implication (i) $\Rightarrow$ (iii) is tricky, although content-free (GOTO (5.9) if you want to avoid a headache): by assumption (i), there are inverse rational maps $f\colon V \dashrightarrow W$ and $g\colon W \dashrightarrow V$; now set $V' = \operatorname{dom} f \subset V$ and $\varphi = f_{|V'}\colon V' \to W$, and similarly $W' = \operatorname{dom} g \subset W$ and $\psi = g_{|W'}\colon W' \to V$. In the diagram

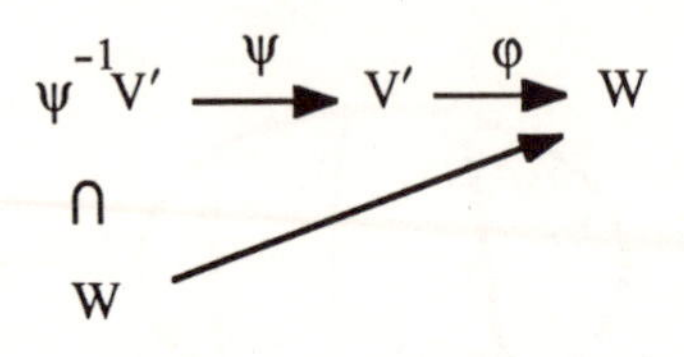

all the arrows are morphism, and $\mathrm{id}_{W|\psi^{-1}V'} = \psi\circ\varphi$ (as morphisms) follows from $\mathrm{id}_W = f\circ g$ (as rational maps). Hence

$$\varphi(\psi(P)) = P \quad \text{for all } P \in \psi^{-1}V'.$$

Now set $V_0 = \varphi^{-1}\psi^{-1}V'$, and $W_0 = \psi^{-1}\varphi^{-1}W'$; then by construction, $\varphi\colon V_0 \to \psi^{-1}V'$ is a morphism. However, $\psi^{-1}V' \subset W_0$, since $P \in \psi^{-1}V'$ implies that $\varphi(\psi(P)) = P$, so that $P \in \psi^{-1}\varphi^{-1}W' = W_0$. Therefore, $\varphi\colon V_0 \to W_0$ is a morphism, and similarly $\psi\colon W_0 \to V_0$. Q.E.D.

**(5.9) Rational varieties.** The notion of birational equivalence discussed in (5.8) is of key importance in algebraic geometry. Condition (iii) in the proposition says that the 'meat' of the varieties V and W is the same, although they may differ a bit around the edges; an example of the use of birational transformations is blowing up a singular variety to obtain a non-singular one, see (6.11) below. An important particular case of Proposition 5.8 is the following result.

**Corollary.** Given a variety V, the following two conditions are equivalent:

(a) the function field $k(V)$ is a purely transcendental extension of k, that is $k(V) \cong k(t_1,.. t_n)$ for some n;

(b) there exists a dense open set $V_0 \subset V$ which is isomorphic to a dense open subset $U_0 \subset \mathbb{A}^n$.

A variety satisfying these conditions is said to be *rational*. Condition (b) is a precise version of the statement that V can be parametrised by n independent variables. This notion has already appeared implicitly several times in these notes (for example, (1.1), (2.1), (3.11, (b)), (5.7, II)). A large proportion of the elementary applications of algebraic geometry to other branches of math are related one way or another to rational varieties.

**(5.10) Reduction to a hypersurface.** An easy consequence of the discussion of Noether normalisation at the end of §3 is that every variety is birational to a hypersurface: firstly, since birational questions only depend on a dense open set, and any open set contains a dense open subset isomorphic to an affine variety (by (4.13)), I only need to consider an affine variety $V \subset \mathbb{A}^n$. It was proved in (3.18) that there exist elements $y_1,.. y_{m+1} \in k[V]$ which generate the field extension $k \subset k(V)$, and such that $y_1,.. y_m$ are algebraically independent, and $y_{m+1}$ is algebraic over $k(y_1,.. y_m)$. These elements thus define a morphism $V \to \mathbb{A}^{m+1}$ which is a birational equivalence of $V$ with a hypersurface $V' \subset \mathbb{A}^{m+1}$.

**(5.11) Products.** If $V$ and $W$ are two affine varieties then there is a natural sense in which $V \times W$ is again a variety: if $V \subset \mathbb{A}^n$ and $W \subset \mathbb{A}^m$ then $V \times W$ is the subset of $\mathbb{A}^{n+m}$ given by

$$\{((\alpha_1,.. \alpha_n); (\beta_1,.. \beta_m)) \mid f(\underline{\alpha}) = 0 \ \forall\, f \in I(V),\ g(\underline{\beta}) = 0 \ \forall\, g \in I(W)\}.$$

It's easy to check that $V \times W$ remains irreducible. Note however that the Zariski topology of the product is not the product of the Zariski topologies (see Ex. 5.10).

The case of projective varieties is not so obvious; to be able to define products, we need to know that $\mathbb{P}^n \times \mathbb{P}^m$ is itself a projective variety. Notice that it is definitely not isomorphic to $\mathbb{P}^{n+m}$ (see Ex. 5.2, (ii)). To do this, I use a construction rather similar in spirit to that of (5.7, I): make an embedding (the 'Segre embedding')

$$\varphi\colon \mathbb{P}^n \times \mathbb{P}^m \longrightarrow S_{n,m} \subset \mathbb{P}^N, \quad \text{where} \quad N = (n+1)(m+1) - 1$$

as follows: $\mathbb{P}^N$ is the projective space with homogeneous coordinates

$$(U_{ij})_{i = 0,..\, n;\ j = 0,..\, m}$$

it's useful to think of the $U_{ij}$ as being set out in a matrix

$$\begin{bmatrix} U_{00} & .. & U_{0m} \\ U_{10} & .. & .. \\ .. & .. & U_{nm} \end{bmatrix}$$

Then define $\varphi$ by $((X_0,.. X_n), (Y_0,.. Y_m)) \mapsto (X_iY_j)_{i = 0,..\, n;\ j = 0,..\, m}$. This is obviously a well-defined morphism, and the image $S_{n,m}$ is easily seen to be the projective subvariety given by

$$\operatorname{rk}\begin{bmatrix} U_{00} & \cdots & U_{0m} \\ U_{10} & \cdots & \cdots \\ \cdots & \cdots & U_{nm} \end{bmatrix} \le 1, \text{ that is, } \det\begin{vmatrix} U_{ik} & U_{i\ell} \\ U_{jk} & U_{j\ell} \end{vmatrix} = 0 \quad \begin{matrix} \forall\ i, j = 0, \ldots n \\ \forall\ k, \ell = 0 \ldots m \end{matrix}.$$

We get an inverse map $S_{n,m} \to \mathbb{P}^n \times \mathbb{P}^m$ as follows. For $P \in S_{n,m}$ there exists at least one pair $(i, j)$ such that $U_{ij}(P) \neq 0$; fixing this $(i, j)$, send

$$S_{n,m} \ni P \mapsto ((U_{0j}, \ldots U_{nj}), (U_{i0}, \ldots U_{im})) \in \mathbb{P}^n \times \mathbb{P}^m.$$

Note that the choice of $(i, j)$ doesn't matter, since the matrix $U_{ij}(P)$ has rank 1, and hence all its rows and all its columns are proportional.

From this it is not hard to see that if $V \subset \mathbb{P}^n$ and $W \subset \mathbb{P}^m$ are projective varieties, then $V \times W \subset \mathbb{P}^n \times \mathbb{P}^m \cong S_{n,m} \subset \mathbb{P}^N$ is again a projective variety (see Ex. 5.11).

## Exercises to §5.

**5.1.** Prove that a regular function on $\mathbb{P}^1$ is a constant. (Hint: use the notation of (5.1); suppose that $f \in k(\mathbb{P}^1)$ is regular at every point of $\mathbb{P}^1$. Apply (4.8, II) to the affine piece $\mathbb{A}^1_{(0)}$, to show that $f = p(x_0) \in k[x_0]$; on the other affine piece $\mathbb{A}^1_{(\infty)}$, $f = p(1/y_1) \in k[y_1]$. Now, how can it happen that $p(1/y_1)$ is a polynomial?). Deduce that there are no non-constant morphisms $\mathbb{P}^1 \to \mathbb{A}^m$ for any m.

**5.2.** (i) Show that the Segre embedding of $\mathbb{P}^1 \times \mathbb{P}^1$ (as in (5.10)) gives an isomorphism of $\mathbb{P}^1 \times \mathbb{P}^1$ with the quadric $S_{1,1} = Q\colon (X_0X_3 = X_1X_2) \subset \mathbb{P}^3$.

(ii) What are the images in Q of the two families of lines $\{p\} \times \mathbb{P}^1$ and $\mathbb{P}^1 \times \{p\}$ in $\mathbb{P}^1 \times \mathbb{P}^1$? Use this to find some disjoint lines in $\mathbb{P}^1 \times \mathbb{P}^1$, and conclude from this that $\mathbb{P}^1 \times \mathbb{P}^1 \not\cong \mathbb{P}^2$.

(The fact that a quadric surface has two rulings by straight lines has applications in civil engineering: if you're trying to build a curved surface out of concrete, it's an obvious advantage to be able to determine the shape of the surface by imposing linear constraints. See [M. Berger, 14.4.6–7 and 15.3.3] for a discussion and pictures.)

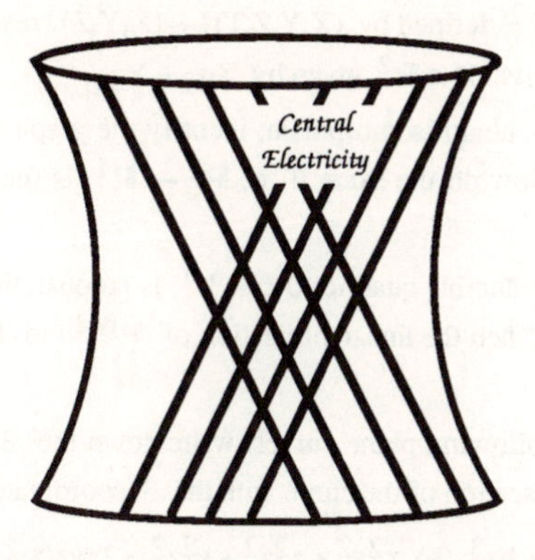


(iii) Show that there are two lines of $Q$ passing through the point $P = (1, 0, 0, 0)$, and that the complement $U$ of these two lines is the image of $\mathbb{A}^1 \times \mathbb{A}^1$ under the Segre embedding.

(iv) Show that under the projection $\pi_{|Q}: Q \dashrightarrow \mathbb{P}^2$, (in the notation of (5.7, (II))), $U$ maps isomorphically to a copy of $\mathbb{A}^2$, and the two lines through $P$ are mapped to points of $\mathbb{P}^2$.

(v) In the notation of (5.7, (II)), find $\operatorname{dom} \pi$ and $\operatorname{dom} \varphi$, and give a geometrical interpretation of the singularities of $\pi$ and $\varphi$.

**5.3.** Which of the following expressions define rational maps $\varphi: \mathbb{P}^n \dashrightarrow \mathbb{P}^m$ between projective spaces of the appropriate dimensions $(n, m = 1$ or $2)$? In each case, determine $\operatorname{dom} \varphi$, say if $\varphi$ is birational, and if so describe the inverse map.

(a) $(x, y, z) \mapsto (x, y)$; (b) $(x, y) \mapsto (x, y, 1)$; (c) $(x, y) \mapsto (x, y, 0)$;

(d) $(x, y, z) \mapsto (1/x, 1/y, 1/z)$; (e) $(x, y, z) \mapsto ((x^3 + y^3)/z^3, y^2/z^2, 1)$;

(f) $(x, y, z) \mapsto (x^2 + y^2, y^2, y^2)$.

**5.4.** The rational normal curve (see (5.7, (I)) of degree 3 is the curve $C \subset \mathbb{P}^3$ defined by the 3 quadrics $C = Q_1 \cap Q_2 \cap Q_3$, where

$$Q_1: (XZ = Y^2),\ Q_2: (XT = YZ),\ Q_3: (YT = Z^2);$$

this curve is also well-known as the twisted cubic, where 'twisted' refers to the fact that it is not a plane curve. Check that for any two of the quadrics $Q_i, Q_j$, the intersection $Q_i \cap Q_j = C \cup \ell_{ij}$, where $\ell_{ij}$ is a certain line. So this curve in 3-space is not the intersection of any 2 of the quadrics.

**5.5.** Let $Q_1: (XZ = Y^2)$ and $F: (XT^2 - 2YZT + Z^3 = 0)$; prove that $C = Q_1 \cap F$ is the twisted cubic curve of Ex. 5.4 (Hint: start by multiplying $F$ by $X$; subtracting a suitable multiple of $Q_1$ this becomes a perfect square).

**5.6.** Let $C \subset \mathbb{P}^3$ be an irreducible curve defined by $C = Q_1 \cap Q_2$, where $Q_1: (TX = q_1), Q_2: (TY = q_2)$, with $q_1, q_2$ quadratic forms in $X, Y, Z$. Show that the projection $\pi: \mathbb{P}^3 \dashrightarrow \mathbb{P}^2$ defined by $(X,Y,Z,T) \mapsto (X,Y,Z)$ restricts to an isomorphism of $C$ with the plane curve $D \subset \mathbb{P}^2$ given by $Xq_2 = Yq_1$.

**5.7.** Let $\varphi: \mathbb{P}^1 \to \mathbb{P}^1$ be an isomorphism; identify the graph of $\varphi$ as a subvariety of $\mathbb{P}^1 \times \mathbb{P}^1 \cong Q \subset \mathbb{P}^3$. Now do the same if $\varphi: \mathbb{P}^1 \to \mathbb{P}^1$ is the 2–to–1 map given by $(X,Y) \mapsto (X^2,Y^2)$.

**5.8.** Prove that any irreducible quadric $Q \subset \mathbb{P}^{n+1}$ is rational, that is, show that if $P \in Q$ is a non-singular point, then the linear projection of $\mathbb{P}^{n+1}$ to $\mathbb{P}^n$ induces a birational map $Q \dashrightarrow \mathbb{P}^{n+1}$.

**5.9.** For each of the following plane curves, write down the 3 standard affine pieces, and determine the intersection of the curve with the 3 coordinate axes:

(a) $y^2z = x^3 + axz^2 + bz^3$; (b) $x^2y^2 + x^2z^2 + y^2z^2 = 2xyz(x + y + z)$;

(c) $xz^3 = (x^2 + z^2)y^2$.

**5.10.** (i) Prove that the product of two irreducible algebraic sets is again irreducible (Hint: the subsets $V \times \{w\}$ are irreducible for $w \in W$; given an expression $V \times W = U_1 \cup U_2$, consider the subsets $W_i = \{w \in W \mid V \times \{w\} \subset U_i\}$ for $i = 1, 2$).

(ii) Describe the closed sets of the topology on $\mathbb{A}^2 = \mathbb{A}^1 \times \mathbb{A}^1$ which is the product of the Zariski topologies on the two factors; now find a closed subset of the Zariski topology of $\mathbb{A}^2$ not of this form.

**5.11.** (a) If $\mathbb{A}^n_{(0)}$ and $\mathbb{A}^m_{(0)}$ are standard affine pieces of $\mathbb{P}^n$ and $\mathbb{P}^m$ respectively, verify that $\mathbb{A}^n_{(0)} \times \mathbb{A}^m_{(0)}$ is mapped isomorphically to an affine piece of the variety $S_{n,m} \subset \mathbb{P}^N$, say $S_{(0)} \subset \mathbb{A}^N$, and that the $N$ coordinates of $\mathbb{A}^N$ restrict to $X_1,.. X_n, Y_1,.. Y_m$ and $nm$ terms $X_iY_j$.

(b) If $V \subset \mathbb{P}^n$ and $W \subset \mathbb{P}^m$, prove that the product $V \times W$ is a projective subvariety of $\mathbb{P}^n \times \mathbb{P}^m = S_{n,m} \subset \mathbb{P}^N$ (Hint: the product of the affine pieces $V_{(0)} \times W_{(0)} \subset \mathbb{A}^{n+m}$ is a subvariety defined by polynomials as explained in (5.10); show that each of these is the restriction to $\mathbb{A}^{n+m} \cong S_{(0)}$ of a homogeneous polynomial in the $U_{ij}$).

**5.12.** Let $C$ be the cubic curve of (5.1); prove that any regular function $f$ on $C$ is constant. Proceed in the following steps:

**Step 1.** Applying (4.8, II) to the affine piece $C_{(0)}$, write $f = p(x, y) \in k[x, y]$.

**Step 2.** Subtracting a suitable multiple of the relation $y^2 - x^3 - ax - b$, assume that $p(x, y) = q(x) + yr(x)$, with $q, r \in k[x]$.

**Step 3.** Applying (4.8, II) to the affine piece $C_{(\infty)}$, we have

$$f = q(x_1/z_1) + (1/z_1)r(x_1/z_1) \in k[C_{(\infty)}],$$

and hence there exists a polynomial $S(x_1, z_1)$ such that

$$q(x_1/z_1) + (1/z_1)r(x_1/z_1) = S(x_1, z_1);$$

**Step 4.** Clearing the denominator, and using the fact that $k[C_{(\infty)}] = k[x_1, z_1]/g$, where $g = z_1 - x_1{}^3 - axz_1{}^2 - bz_1{}^3$, deduce a polynomial identity

$$Q_m(x_1, z_1) + R_{m-1}(x_1, z_1) \equiv S(x_1, z_1)z_1{}^m + A(x_1, z_1)g$$

in $k[x_1, z_1]$.

**Step 5.** Now if we write $S = S^+ + S^-$ and $A = A^+ + A^-$ for the decomposition into terms of even and odd degree, and note that $g$ has only terms of odd degree, this identity splits into two:

$$Q_m \equiv S^+z_1{}^m + A^-g \text{ and } R_{m-1} \equiv S^-z_1{}^m + A^+g$$

if $m$ is even, and an analogous expression if $m$ is odd.

**Step 6.** $Q_m$ is homogeneous of degree $m$, and hence $A^-g$ has degree $\geq m$; by considering the term of least degree in $A^-g$, prove that $Q_m$ is divisible by $z_1$. Similarly for $z_1$. By taking the minimum value of $z_1$ in the identity of Step 4, prove that $q(x)$ has degree $0$ and $r(x) = 0$.

**5.13. Veronese surface.** Study the embedding $\varphi: \mathbb{P}^2 \to \mathbb{P}^5$ given by $(X, Y, Z) \mapsto (X^2, XY, XZ, Y^2, YZ, Z^2)$; write down the equations defining the image $S = \varphi(\mathbb{P}^2)$, and prove that $\varphi$ is an isomorphism (by writing down the equations of the inverse morphism). Prove that the lines of $\mathbb{P}^2$ go over into conics of $\mathbb{P}^5$, and that conics of $\mathbb{P}^2$ go over into twisted quartics of $\mathbb{P}^5$ (see (5.7)).

For any line $\ell \subset \mathbb{P}^2$, write $\pi(\ell) \subset \mathbb{P}^5$ for the projective plane spanned by the conic $\varphi(\ell)$. Prove that the union of $\pi(\ell)$ taken over all $\ell \subset \mathbb{P}^2$ is a cubic hypersurface $\Sigma \subset \mathbb{P}^5$. (Hint: as in (5.7) and (5.11), you can write the equations defining $S$ in the form rank $M \leq 1$, where $M$ is a symmetric $3 \times 3$ matrix with entries the 6 coordinates of $\mathbb{P}^5$; then show that $\Sigma$: $(\det M = 0)$. See [Semple and Roth, p.128] for more details.)

# §6. Non-singularity and dimension

**(6.1) Non-singular points of a hypersurface.** Suppose $f \in k[X_1,\dots X_n]$ is irreducible, $f \notin k$, and set $V = V(f) \subset \mathbb{A}^n$; let $P = (a_1,\dots a_n) \in V$, and $\ell$ be a line through P. Since $P \in V$, obviously P is a root of $f_{|\ell}$.
**Question:** When is P a multiple root of $f_{|\ell}$?
**Answer:** If and only if $\ell$ is contained in the linear subspace

$$T_PV\colon \left(\sum_i \frac{\partial f}{\partial X_i}(P)\cdot(X_i - a_i) = 0\right) \subset \mathbb{A}^n,$$

called the *tangent space* to V at P.

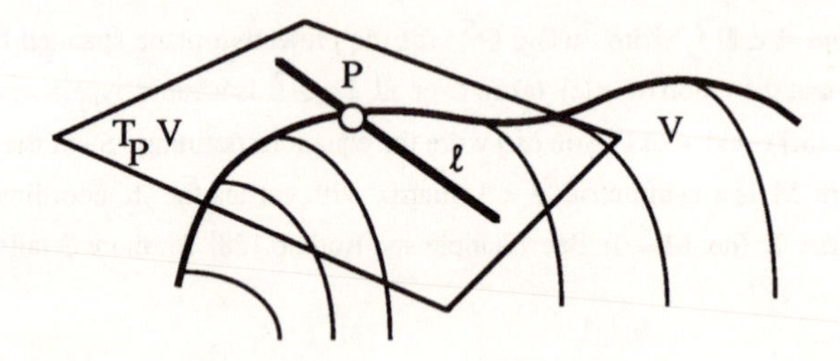


To prove this, parametrise $\ell$ as

$$\ell\colon X_i = a_i + b_iT,$$

where $P = (a_1,\dots a_n)$ and $(b_1,\dots b_n)$ is the slope or direction vector of $\ell$. Then $f_{|\ell} = f(,\dots a_i + b_iT,\dots) = g(T)$ is a polynomial in T, and we know that $(T = 0)$ is one root of g. Hence

$$0 \text{ is a multiple root of } g \iff \frac{\partial g}{\partial T}(0) = 0,$$

that is,

$$\iff \sum_i b_i \frac{\partial f}{\partial X_j}(P) = 0 \iff \ell \subseteq T_PV.$$

**Definition.** $P \in V \subset \mathbb{A}^n$ is a *non-singular point* of V if $\partial f/\partial X_i(P) \neq 0$ for some i; otherwise P is a *singular point*, or a *singularity* of V.

Obviously $T_PV$ is an $(n-1)$-dimensional affine subspace of $\mathbb{A}^n$ if P is non-singular, and $T_PV = \mathbb{A}^n$ if $P \in V$ is singular.

**(6.2) Remarks.** (a) The derivatives $\partial f/\partial X_i(P)$ appearing above are formal algebraic operations (that is, $\partial/\partial X_i$ takes $X_i^n$ into $nX_i^{n-1}$); no calculus is involved.

(b) Suppose $k = \mathbb{R}$ or $\mathbb{C}$, and that $\partial f/\partial X_i(P) \neq 0$; for clarity let me take $i = 1$. Then the map $p: \mathbb{A}^n \to \mathbb{A}^n$ defined by $(X_1,.. X_n) \mapsto (f, X_2,.. X_n)$ has non-vanishing Jacobian determinant at P, so that by the inverse function theorem, there exists a neighbourhood $P \in U \subset \mathbb{A}^n$ such that $p_{|U}: U \to p(U) \subset \mathbb{A}^n$ is a diffeomorphism of the neighbourhood U with an open set p(U) of $\mathbb{A}^n$ (in the usual topology of $\mathbb{R}^n$ or $\mathbb{C}^n$); that is, $p_{|U}$ is bijective, and both p and $p^{-1}$ are differentiable functions of (real or complex) variables. In other words, $(f, X_2,.. X_n)$ form a new differentiable coordinate system on $\mathbb{A}^n$ near P; this implies that a neighbourhood of P in V: $(f = 0)$ is diffeomorphic to an open set in $\mathbb{A}^{n-1}$ with coordinates $(X_2,.. X_n)$. Thus near a non-singular point P, V is a *manifold* with $(X_2,.. X_n)$ as local parameters.

**(6.3) Proposition.** $V_{\text{non-sing}} = \{P \in V \mid P \text{ is non-singular}\}$ is a dense open set of V for the Zariski topology.

**Proof.** The complement of $V_{\text{non-sing}}$ is the set $V_{\text{sing}}$ of singular points, which is defined by $\partial f/\partial X_i(P) = 0$ for all i, that is

$$V_{\text{sing}} = V(f, \frac{\partial f}{\partial X_1}, .. \frac{\partial f}{\partial X_n}) \subset \mathbb{A}^n,$$

which is closed by definition of the Zariski topology. Since V is irreducible (by (3.11)), to show that the open $V_{\text{non-sing}}$ is dense, I only have to show it's non-empty (by Proposition 4.2); arguing by contradiction, suppose that it's empty, that is, suppose $V = V(f) = V_{\text{sing}}$. Then each of the polynomials $\partial f/\partial X_i$ must vanish on V, therefore (by (3.11) once more) they must be divisible by f in $k[X_1,.. X_n]$; but viewed as a polynomial in $X_i$, $\partial f/\partial X_i$ has degree strictly smaller than f, so that f divides $\partial f/\partial X_i$ implies that in fact $\partial f/\partial X_i = 0$ as a polynomial. Over $\mathbb{C}$, this is obviously only possible if $X_i$ does not appear in f, and if this happens for all i then $f = \text{const.} \in \mathbb{C}$, which is excluded. Over a general field k, $\partial f/\partial X_i = 0$ is only possible if f is an inseparable polynomial in $X_i$, that is, char $k = p$, and $X_i$ only appear in f as the pth power $X_i^p$. If this happens for each i, then by the argument given in (3.16), f is a pth power in $k[X_1,.. X_n]$; this contradicts the fact that f is irreducible. Q.E.D.

**(6.4) Tangent space.**

**Definition.** Let $V \subset \mathbb{A}^n$ be a subvariety, with $V \ni P = (a_1,.. a_n)$. For any $f \in k[X_1,.. X_n]$, write

$$f_P^{(1)} = \sum_i \frac{\partial f}{\partial X_i}(P)\cdot(X_i - a_i).$$

This is an affine linear polynomial (that is, linear plus constant), the 'first-order part' of $f$ at $P$. Now define the *tangent space* to $V$ at $P$ by

$$T_P V = \bigcap (f_P^{(1)} = 0) \subset \mathbb{A}^n,$$

where the intersection takes place over all $f \in I(V)$.

**(6.5) Proposition.** The function $V \to \mathbb{N}$ defined by $P \mapsto \dim T_P V$ is an upper semicontinuous function (in the Zariski topology of $V$). In other words, for any integer $r$, the subset

$$S(r) = \{P \in V \mid \dim T_P V \geq r\} \subset V$$

is closed.

**Proof.** Let $(f_1,.. f_m)$ be a set of generators of $I(V)$; it is easy to see that for any $g \in I(V)$, the linear part $g_P^{(1)}$ of $g$ is a linear combination of those of the $f_i$, so that the definition of $T_P V$ simplifies to

$$T_P V = \bigcap_{i=1}^{m} (f_{i,P}^{(1)} = 0) \subset \mathbb{A}^n.$$

Then by elementary linear algebra,

$$P \in S(r) \iff \text{the matrix } \left(\frac{\partial f_i}{\partial X_j}(P)\right)_{i = 1,.. m,\, j = 1,.. n} \text{ has rank } \leq n - r$$

$$\iff \text{every } (n - r + 1) \times (n - r + 1) \text{ minor of } \left(\frac{\partial f_i}{\partial X_j}(P)\right)_{i,j} \text{ vanishes.}$$

Now each entry $\partial f_i/\partial X_j(P)$ of the matrix is a polynomial function of $P$; thus each minor is a determinant of a matrix of polynomials, and so is itself a polynomial. Hence $S(r) \subset V \subset \mathbb{A}^n$ is an algebraic subset. Q.E.D.

**(6.6) Corollary-Definition.** There exists an integer $r$ and a dense open subset $V_0 \subset V$ such that

$$\dim T_P V = r \text{ for } P \in V_0, \text{ and } \dim T_P V \geq r \text{ for all } P \in V.$$

Define $r$ to be the *dimension* of $V$, $\dim V = r$; and say that $P \in V$ is *non-singular* if $\dim T_P V = r$, and *singular* if $\dim T_P V > r$. A variety $V$ is *non-singular* if it is non-singular at each point $P \in V$.

**Proof.** Let $r = \min\{\dim T_P V\}$, taken over all points $P \in V$. Then clearly

$$S(r-1) = \emptyset, \ S(r) = V, \text{ and } S(r+1) \subsetneq V;$$

therefore $S(r) \setminus S(r+1) = \{P \in V \mid \dim T_P V = r\}$ is open and non-empty. Q.E.D.

**(6.7)** It follows from (6.3) that if $V = V(f) \subset \mathbb{A}^n$ is a hypersurface defined by some non-constant polynomial $f$, then $\dim V = n-1$. On the other hand, for a hypersurface, $k[V] = k[X_1,.. X_n]/(f)$, so that, assuming that $f$ involves $X_1$ in a non-trivial way, the function field of $V$ is of the form

$$k(V) = k(X_2,.. X_n)[X_1]/(f),$$

that is, it is built up from $k$ by adjoining $n-1$ algebraically independent elements, then making an primitive algebraic extension.

**Definition.** If $k \subset K$ is a field extension, the *transcendence degree* of $K$ over $k$ is the maximum number of elements of $K$ algebraically independent over $k$. It is denoted $\operatorname{tr}\deg_k K$.

The elementary theory of transcendence degree of a field extension $K/k$ is formally quite similar to that of the dimension of a vector space: given $\alpha_1,.. \alpha_m \in K$, we know what it means for them to be *algebraically independent* over $k$ (see (3.13)); they *span* the transcendental part of the extension if $K/k(\alpha_1,.. \alpha_m)$ is algebraic; and they form a *transcendence basis* if they are algebraically independent and span. Then it is an easy theorem that a transcendence basis is a maximal algebraically independent set, and a minimal spanning set, and that any two transcendence bases of $K/k$ have the same number of elements (see Ex. 6.1).

Thus for a hypersurface $V \subset \mathbb{A}^n$, $\dim V = n - 1 = \operatorname{tr}\deg_k k(V)$. The rest of this section is concerned with proving that the equality $\dim V = \operatorname{tr}\deg_k k(V)$ holds for all varieties, by reducing to the case of a hypersurface. The first thing to show is that for a point $P \in V$ of a variety, the tangent space $T_P V$, which so far has been discussed in terms of a particular coordinate system in the ambient space $\mathbb{A}^n$, is in fact an intrinsic property of a neighbourhood of $P \in V$.

**(6.7) Intrinsic nature of $T_PV$.**

From now on, given $P = (a_1, .. a_n) \in V \subset \mathbb{A}^n$, I take new coordinates $X_i' = X_i - a_i$ to bring $P$ to the origin, and thus assume that $P = (0, .. 0)$. Then $T_PV \subset \mathbb{A}^n$ is a vector subspace of $k^n$.

**Notation.** Write $m_P$ = ideal of $P$ in $k[V]$, and

$$M_P = \text{ideal } (X_1, .. X_n) \subset k[X_1, .. X_n].$$

Then of course $m_P = M_P/I(V) \subset k[V]$.

**Theorem.** In the above notation,

(a) there is a natural isomorphism of vector spaces

$$(T_PV)^* = m_P/m_P^2,$$

where $(\ )^*$ denotes the dual of a vector space.

(b) If $f \in k[V]$ is such that $f(P) \neq 0$, and $V_f \subset V$ is the standard affine open as in (4.13), then the natural map

$$T_P(V_f) \to T_PV$$

is an isomorphism.

**Proof of (a).** Write $(k^n)^*$ for the vector space of linear forms on $k^n$; this is the vector space with basis $X_1, .. X_n$. Since $P = (0, .. 0)$, for any $f \in k[X_1, .. X_n]$, the linear part $f_P^{(1)}$ is naturally an element of $(k^n)^*$; define a map $d\colon M_P \to (k^n)^*$ by taking $f \in M_P$ into $df = f_P^{(1)}$.

Now $d$ is surjective, since $X_i \in M_P$ go into the natural basis of $(k^n)^*$; also $\ker d = M_P^2$, since

$$f_P^{(1)} = 0 \iff f \text{ starts with quadratic terms in } X_1, .. X_n \iff f \in M_P^2.$$

Hence $M_P/M_P^2 \cong (k^n)^*$. This is statement (a) for the special case $V = \mathbb{A}^n$. In the general case, dual to the inclusion $T_PV \subset k^n$, there is a restriction map $(k^n)^* \to (T_PV)^*$, taking a linear form $\lambda$ on $k^n$ into its restriction to $T_PV$; composing then defines a map

$$D\colon M_P \to (k^n)^* \to (T_PV)^*.$$

The composite $D$ is surjective since each factor is. I claim that the kernel of $D$ is just $M_P^2 + I(V)$, so that

$$m_P/m_P^2 = M_P/(M_P^2 + I(V)) \cong (T_PV)^*,$$

as required. To prove the claim,

$$f \in \ker D \iff f_P^{(1)}|T_PV = 0$$

$$\iff f_P^{(1)} = \sum_i a_i g_{i,P}^{(1)} \quad \text{for some } g_i \in I(V)$$

(since $T_PV \subset k^n$ is the vector subspace defined by $(g_P^{(1)} = 0)$ for $g \in I(V)$)

$$\iff f - \sum_i a_i g_i \in M_P^2 \text{ for some } g_i \in I(V) \iff f \in M_P^2 + I(V). \qquad \text{Q.E.D.}$$

The proof of (b) of Proposition 6.7 is left to the reader (see Ex. 6.2).

**(6.8) Corollary.** $T_PV$ only depends on a neighbourhood of $P \in V$ up to isomorphism. More precisely, if $P \in V_0 \subset V$ and $Q \in W_0 \subset W$ are open subsets of affine varieties, and $\varphi: V_0 \to W_0$ an isomorphism taking $P$ into $Q$, there is a natural isomorphism $T_PV_0 \to T_QW_0$; hence $\dim T_PV_0 = \dim T_QW_0$.

In particular, if $V$ and $W$ are birationally equivalent varieties then $\dim V = \dim W$.

**Proof.** By passing to a smaller neighbourhood of $P$ in $V$, I can assume $V_0$ is isomorphic to an affine variety (Proposition 4.13). Then so is $W_0$, and $\varphi$ induces an isomorphism $k[V_0] \cong k[W_0]$ taking $m_P$ into $m_Q$. The final sentence holds because by (5.8), $V$ and $W$ contain dense open subsets which are isomorphic.

**(6.9) Theorem.** For any variety $V$, $\dim V = \operatorname{tr deg} k(V)$.

**Proof.** This is known if $V$ is a hypersurface. On the other hand, every variety is birational to a hypersurface (by (5.10)), and both sides of the required relation are the same for birationally equivalent varieties. Q.E.D.

**(6.10) Non-singularity and projective varieties.** Although the above results were discussed in terms of affine varieties, the idea of non-singularity and of dimension applies directly to any variety $V$: a point $P \in V$ is non-singular if it is a non-singular point of an affine open $V_0 \subset V$ containing it; by (6.8), this notion does not depend on the choice of $V_0$. On the other hand, for a projective variety $V \subset \mathbb{P}^n$, it is sometimes useful to consider the tangent space to $V$ at $P$ as a projective subspace of $\mathbb{P}^n$. I give the definition for a hypersurface only: if $V = V(f)$ is a hypersurface defined by a form (= homogeneous polynomial) $f \in k[X_0,.. X_n]$ of degree $d$, and $V \ni P = (a_0,\dots a_n)$, then $\sum \partial f/\partial X_i(P)\cdot X_i = 0$ is the equation of a hyperplane in $\mathbb{P}^n$ which plays the role of the tangent plane to $V$ at $P$. If $P \in \mathbb{A}^n_{(0)}$, then this projective hyperplane is the projective closure of the affine tangent hyperplane to $V_{(0)}$ at $P$, as can be checked easily using Euler's formula:

$$\sum X_i\cdot(\partial f/\partial X_i) = nf \quad \text{if} \quad f \in k[X_0,\dots X_n] \text{ is homogeneous of degree } n.$$

Because of this formula, to find out whether a point $P \in \mathbb{P}^n$ is a singular point of $V$, we only have to check $(n+1)$ out of the $(n+2)$ conditions

$$f(P) = 0,\ \partial f/\partial X_i(P) = 0 \quad \text{for } i = 0,\dots n,$$

so that for example, if the degree of $f$ is not divisible by char $k$,

$$\partial f/\partial X_i(P) = 0 \text{ for } i = 0,\dots n \implies f(P) = 0, \text{ and } P \in V \text{ is a singularity.}$$

**(6.11) Worked example: blow-up.** Let $B = \mathbb{A}^2$ with coordinates $(u, v)$, and $\sigma\colon B \to \mathbb{A}^2$ the map $(u, v) \mapsto (x = u, y = uv)$; clearly, $\sigma$ is a birational morphism: it contracts the $v$-axis $\ell : (u = 0)$ to the origin $0$ and is an isomorphism outside this exceptional set. Let's find out what happens under $\sigma$ to a curve $C : (f = 0) \subset \mathbb{A}^2$; the question will only be of interest if $C$ passes through $0$.

Clearly $\sigma^{-1}(C) \subset B$ is the algebraic subset defined by $(f\circ\sigma)(u, v) = f(u, uv) = 0$; since $0 \in C$ by assumption, it follows that $\ell : (u = 0)$ is contained in $\sigma^{-1}(C)$, or equivalently, that $u \mid f(u, uv)$. It's easy to see that the highest power $u^m$ of $u$ dividing $f(u, uv)$ is equal to the smallest degree $m = a + b$ of the monomials $x^a y^b$ occuring in $f$, that is, the *multiplicity* of $f$ at $0$; so $\sigma^{-1}(C)$ decomposes as the union of the exceptional curve $\sigma^{-1}(0) = \ell$ (with multiplicity $m$), together with a new curve $C_1$ defined by $f_1(u, v) = f(u, uv)/u^m$. Consider the examples

$$\text{(a) } f = \alpha x - y + \dots; \text{ or (b) } f = y^2 - x^2 + \dots; \text{ or (c) } f = y^2 - x^3,$$

where .. denotes terms of higher degree. Clearly in (a) $f$ has multiplicity 1, and $f_1 = \alpha - v + \dots$(where .. consists of terms divisible by $u$), so $C_1$ is again nonsingular, and meets $\ell$ transversally at $(0, \alpha)$; thus $\sigma$ replaces $0 \in \mathbb{A}^2$ with the line $\ell$ whose points correspond to tangent directions at $0$ (excluding $(x = 0)$). In (b) $f_1 = v^2 - 1 + \dots$, so $C_1$ has two nonsingular points $(0, \pm 1)$ above $0 \in C$; thus the blow-up $\sigma$ 'separates the two branches' of the singular curve $C$. In (c) $f_1 = v^2 - u$, so that $C_1$ is nonsingular, but above $0$ it is tangent to the contracted curve $\ell$.

In either case (b) or (c), $\sigma$ replaces a singular curve $C$ by a nonsingular one $C_1$ birational to $C$ (by introducing 'new coordinates' $u = x$, $v = y/x$). This is what is meant by a *resolution of singularities*. In the case of plane curves, a resolution can always be obtained by a chain of blow-ups (see Ex. 6.6 for examples, and [Fulton, p.162-171] for more details), and the process of resolution gives detailed information about the singularities. A crucial theoretical fact is a famous theorem of H. Hironaka that guarantees the possibility of resolving singularities by blow-ups (in any dimension, over a field of characteristic zero); however, the actual process of resolution by blow-ups is in general extremely complicated, and does not necessarily contribute very much to the understanding of individual singularities.

## Exercises to §6.

**6.1.** Let $k \subset K$ be a field extension, and $(u_1,.. u_r)$, $(v_1,.. v_s)$ two sets of elements of $K$; suppose that $(u_1,.. u_r)$ are algebraically independent, and that $(v_1,.. v_s)$ span the extension $k \subset K$ algebraically. Prove that $r \leq s$. (Hint: the inductive step consists of assuming that $(u_1,.. u_i, v_{i+1},.. v_s)$ span $K/k$ algebraically, and considering $u_{i+1}$.) Deduce that any two transcendence bases of $K/k$ have the same number of elements.

**6.2.** Prove Proposition 6.7, (b). (Hint: $I(V_f) = (I(V), Yf - 1) \subset k[X_1,.. X_n, Y]$, so that if $Q = (a_1,.. a_n, b) \in V_f$, then $T_Q V_f \subset \mathbb{A}^{n+1}$ is defined by the equations for $T_P V \subset \mathbb{A}^n$, together with one equation involving $Y$.)

**6.3.** Determine all the singular points of the following curves in $\mathbb{A}^2$.

(a) $y^2 = x^3 - x$; (b) $y^2 = x^3 - 6x^2 + 9x$;

(c) $x^2y^2 + x^2 + y^2 + 2xy(x + y + 1) = 0$; (d) $x^2 = x^4 + y^4$;

(e) $xy = x^6 + y^6$; (f) $x^3 = y^2 + x^4 + y^4$; (g) $x^2y + xy^2 = x^4 + y^4$.

**6.4.** Find all the singular points of the surfaces in $\mathbb{A}^3$ given by

(a) $xy^2 = z^2$; (b) $x^2 + y^2 = z^2$; (c) $xy + x^3 + y^3 = 0$.

(You will find it useful to sketch the real parts of the surfaces, to the limits of your ability; algebraic geometers usually can't draw.)

**6.5.** Show that the hypersurface $X_d \subset \mathbb{P}^n$ defined by

$$X_0^d + X_1^d + .. X_n^d = 0$$

is non-singular (if $\operatorname{char} k$ does not divide $d$).

**6.6.** (a) Let $C_n \subset \mathbb{A}^2$ be the curve given by $f_n\colon y^2 - x^{2n+1}$ and $\sigma\colon B \to \mathbb{A}^2$ be as in (6.11), with $\ell = \sigma^{-1}(0)$; show that $\sigma^{-1}(C_n)$ decomposes as the union of $\ell$ together with a curve isomorphic to $C_{n-1}$. Deduce that $C_n$ can be resolved by a chain of $n$ blow-ups.

(b) Show how to resolve the following curve singularities by making one or more blow-ups: (i) $y^3 = x^4$; (ii) $y^3 = x^5$; (iii) $(y^2 - x^2)(y^2 - x^5) = x^8$.

**6.7.** Prove that the intersection of a hypersurface $V \in \mathbb{A}^n$ (not a hyperplane) with the tangent hyperplane $T_P V$ is singular at $P$.

# §7. The 27 lines on a cubic surface

In this section $S \subset \mathbb{P}^3$ will be a non-singular cubic surface, given by a homogeneous cubic $f = f(X, Y, Z, T)$. Consider the lines $\ell$ of $\mathbb{P}^3$ lying on $S$.

**(7.1) Consequences of non-singularity.**
**Proposition.** (a) There exists at most 3 lines of $S$ through any point $P \in S$; if there are 2 or 3, they must be coplanar. The picture is:

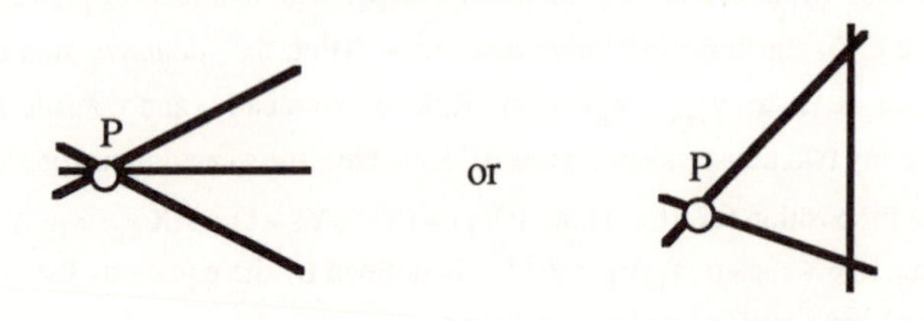


(b) Every plane $\Pi \subset \mathbb{P}^3$ intersects $S$ in one of the following: (i) an irreducible cubic; or (ii) a conic plus a line; or (iii) 3 distinct lines.

**Proof.** (a) If $\ell \subset S$ then $\ell = T_P\ell \subset T_PS$, so that all lines of $S$ through $P$ are contained in the plane $T_PS$; there are at most 3 of them by (b).

(b) I have to prove that a multiple line is impossible: if $\Pi$: $(T = 0)$ and $\ell$: $(Z = 0) \subset \Pi$, then to say that $\ell$ is a multiple line of $S \cap \Pi$ means that $f$ is of the form

$$f = Z^2 \cdot A(X, Y, Z, T) + T \cdot B(X, Y, Z, T),$$

with $A$ a linear form, $B$ a quadratic form. Then $S$: $(f = 0)$ is singular at a point where $Z = T = B = 0$; this is a non-empty set, since it is the set of roots of $B$ on the line $\ell$: $(Z = T = 0)$.

**(7.2) Proposition.** There exists at least one line $\ell$ on $S$.

There are several approaches to proving this. A standard argument is by a dimension-count: lines of $\mathbb{P}^3$ are parametrised by a 4-dimensional variety, and for a line $\ell$ to lie on $S$ imposes 4 conditions on $\ell$ (because the restriction of $f$ to $\ell$ is a cubic form, the 4 coefficients of which must vanish). A little work is needed to turn this into a rigorous proof, since a priori it shows only that the set of lines has dimension $\geq 0$, and not that it is non-empty (see high-brow notes (8.15) for a discussion of the traditional proof and the difficulties involved in it).

It is also perfectly logical to assume the proposition (restrict attention only to cubic surfaces containing lines). There follows a direct proof of (7.2) by coordinate geometry and elimination. If you prefer to omit it, GOTO (7.3). The proof divides up into 3 steps.

**Step 1** (Preliminary construction). In suitable coordinates,

$$f = X^2 Z_0 - Y_0{}^3 + gT$$

where $g = g_2(X, Y_0, Z_0, T)$ is a quadratic form; by non-singularity of $S$ at $(1,0,0,0)$, the coefficient of $Z_0{}^2$ in $g$ is

$$g(0, 0, 1, 0) \neq 0.$$

**Proof.** It is not hard to see that there exists a line $\ell \subset \mathbb{P}^3$ such that $f_{|\ell}$ has a triple root at a point $P \in \ell$; the tangent plane $T_P S$ then intersects $S$ in a cuspidal cubic. Details are left as an exercise (see Ex. 7.3)).

**Step 2** (Expand $f$ in powers of $X$). Consider the variable point

$$P_\alpha = (1, \alpha, \alpha^3, 0) \in S,$$

and choose new coordinates $(X, Y, Z, T)$ with

$$Y = Y_0 - \alpha X \quad \text{and} \quad Z = Z_0 - \alpha^3 X.$$

so that $P_\alpha = (1,0,0,0)$. Write $\gamma$ for the bilinear form associated to $g$; then $g(X, Y + \alpha X, Z + \alpha^3 X, T)$ can be expanded out in powers of $X$:

$$g(X, Y + \alpha X, Z + \alpha^3 X, T) =$$
$$= g(1, \alpha, \alpha^3, 0)X^2 + 2\gamma(1, \alpha, \alpha^3, 0; 0, Y, Z, T)X + g(0, Y, Z, T).$$

It follows that

$$f = X^2 \ell + Xm + n,$$

where $\ell, m$ and $n$ are forms in $(Y, Z, T)$ of degrees 1, 2 and 3, given by

$$\ell = Z - 3\alpha^2 Y + g(1, \alpha, \alpha^3, 0)T$$

$$m = -3\alpha Y^2 + 2\gamma(1, \alpha, \alpha^3, 0; 0, Y, Z, T)T$$

$$n = -Y^3 + g(0, Y, Z, T)T.$$

**Step 3** (Main claim). There exists a polynomial $\varphi = \varphi_{27}(\alpha)$ which is monic of degree 27 in $\alpha$ such that

$$\varphi(\alpha) = 0 \iff \ell = m = n = 0 \text{ have a common zero } (\eta:\zeta:\tau) \text{ in } \mathbb{P}^2.$$

**Proof.** This is a standard elimination calculation: $\ell$ gives $Z$ as a linear form in $Y$ and $T$; substituting in $m$ and $n$ gives quadratic and cubic forms in $Y, T$

$$\mu(Y,T) = a_0Y^2 + a_1YT + a_2T^2,$$

$$\nu(Y, T) = b_0Y^3 + b_1Y^2T + b_2YT^2 + b_3T^3,$$

whose coefficients $a_0, .. b_3$ are polynomials in $\alpha$. Now by Ex. 1.10, $\mu$ and $\nu$ have a common zero $(\eta:\tau)$ if and only if a certain determinant involving the coefficients $a_0, .. b_3$ vanishes. The existence of $\varphi$ follows from this, and the fact that $\deg \varphi \le 27$ comes by looking more closely at the same argument; to see that $\varphi$ is monic is more fun.

To simplify notation, write $a^{(d)}$, $b^{(d)}$, etc. to denote monic polynomials of degree $d$ in $\alpha$; as observed in Step 1, $g(0, 0, 1, 0) \neq 0$, so that I can suppose that $g(0,0,1,0) = 1$. Obviously then, $g(1, \alpha, \alpha^3, 0) = b^{(6)}$ is monic of degree 6 in $\alpha$.

To start the elimination, substitute $Z = 3\alpha^2Y - b^{(6)}T$ in $m$:

$$m = -3\alpha Y^2 + 2\gamma(1, \alpha, \alpha^3, 0; 0, Y, 3\alpha^2Y - b^{(6)}T, T)T;$$

now observe that $\gamma(1, \alpha, \alpha^3, 0; 0, Y, 3\alpha^2Y - b^{(6)}T, T)$ is of the form

$$3c^{(5)}Y - d^{(9)}T.$$

(Because on expanding out the bilinear form, given that $\gamma(0, 0, 1, 0; 0, 0, 1, 0) = 1$, the highest power of $\alpha$ in the coefficients of $Y$ and $T$ comes from $\alpha^3$ against $3\alpha^2Y - b^{(6)}T$.) Therefore $m$ is of the form

$$m = -3\alpha Y^2 + 6c^{(5)}Y - 2d^{(9)};$$

similarly, the leading term of $n = -Y^3 + g(0, Y, 3\alpha^2Y - b^{(6)}T, T)T$ comes from the squared term in $3\alpha^2Y - b^{(6)}T$, so that

$$n = -Y^3 + 9e^{(4)}Y^2T - 6f^{(8)}YT^2 + g^{(12)}T^3.$$

Now by the result of Ex. 1.10, $m$ and $n$ have a common root if and only if $\varphi(\alpha) = 0$, where

$$\varphi(\alpha) = \det \begin{vmatrix} -3\alpha & 6c^{(5)} & -2d^{(9)} & & \\ & -3\alpha & 6c^{(5)} & -2d^{(9)} & \\ & & -3\alpha & 6c^{(5)} & -2d^{(9)} \\ -1 & 9e^{(4)} & -6f^{(8)} & g^{(12)} & \\ & -1 & 9e^{(4)} & -6f^{(8)} & g^{(12)} \end{vmatrix}.$$

It is not hard to see that $\varphi(\alpha)$ is of degree $\leq 27$; to see that it is monic, note that its leading term comes from taking the leading term in each entry of the determinant:

$$\det \begin{vmatrix} -3\alpha & 6\alpha^5 & -2\alpha^9 & & \\ & -3\alpha & 6\alpha^5 & -2\alpha^9 & \\ & & -3\alpha & 6\alpha^5 & -2\alpha^9 \\ -1 & 9\alpha^4 & -6\alpha^8 & \alpha^{12} & \\ & -1 & 9\alpha^4 & -6\alpha^8 & \alpha^{12} \end{vmatrix} = \alpha^{27}\cdot\det \begin{vmatrix} -3 & 6 & -2 & & \\ & -3 & 6 & -2 & \\ & & -3 & 6 & -2 \\ -1 & 9 & -6 & 1 & \\ & -1 & 9 & -6 & 1 \end{vmatrix} = \alpha^{27}.$$

The statement in Step 3 proves (7.2). Indeed, let $\alpha$ be a root of $\varphi_{27}$, so that $\ell, m, n$ have a common root $(\eta : \zeta : \tau)$; then since

$$S: (f = X^2\ell + Xm + n = 0) \subset \mathbb{P}^3,$$

it follows that the point

$$(1, 0, 0, 0) + \lambda(0, \eta, \zeta, \tau)$$

belongs to $S$ for each value of $\lambda$. This gives a line on $S$. Q.E.D.

**(7.3) Proposition.** Given a line $\ell \subset S$, there exists exactly 5 pairs $(\ell_i, \ell_i')$ of lines of $S$ meeting $\ell$, in such a way that

(i) for $i = 1,.. 5$, $\ell \cup \ell_i \cup \ell_i'$ is coplanar,

and (ii) for $i \neq j$, $(\ell_i \cup \ell_i') \cap (\ell_j \cup \ell_j') = \varnothing$.

**Proof** (taken from [Beauville, p.51]). If $\Pi$ is a plane (that is, $\Pi = \mathbb{P}^2 \subset \mathbb{P}^3$), then $\Pi \cap S = \ell +$ conic, (since $f_{|\Pi}$ is divisible by the equation of $\ell$). This conic can either be singular or non-singular:

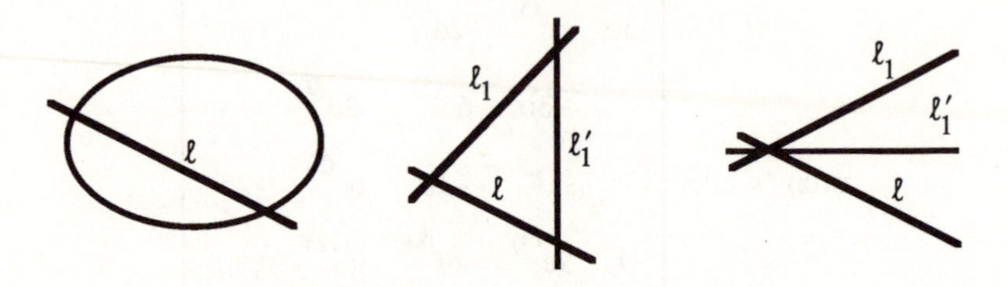


I have to prove that there are exactly 5 distinct planes $\Pi_i \supset \ell$ for which the singular case occurs. The fact, stated as property (ii) that lines in different planes are disjoint, will then follow from (7.1), (a).

Suppose that $\ell$: $(Z = T = 0)$; then I can expand f out as

$$f = AX^2 + 2BXY + CY^2 + 2DX + 2EY + F, \qquad (*)$$

where A, B, C, D, E, F $\in k[Z, T]$, with A, B and C linear forms, D and E quadratic forms, and F a cubic form. If I consider this equation as a variable conic in X and Y, it's singular if and only if

$$\Delta(Z, T) = \det \begin{vmatrix} A & B & D \\ B & C & E \\ D & E & F \end{vmatrix} = 0.$$

To be more precise, any plane through $\ell$ is given by $\Pi$: $(\mu Z = \lambda T)$; if $\mu \neq 0$, I can assume $\mu = 1$, so that $Z = \lambda T$. Then in terms of the homogenous coordinates (X, Y, T) on $\Pi$, $f_{|\Pi} = T \cdot Q(X, Y, T)$, where

$$Q = A(\lambda, 1)X^2 + 2B(\lambda, 1)XY + C(\lambda, 1)Y^2$$
$$+ 2D(\lambda, 1)TX + 2E(\lambda, 1)TY + F(\lambda, 1)T^2.$$

Now $\Delta(Z, T)$ is a homogeneous quintic, so by (1.8), it has 5 roots counted with multiplicities. To prove the proposition, I have to show that it doesn't have multiple roots; this also is a consequence of the non-singularity of S.

**Claim.** $\Delta(Z, T)$ has only simple roots.

Suppose $(Z = 0)$ is a root of $\Delta$, and let $\Pi$: $(Z = 0)$ be the corresponding plane; I have to prove that $\Delta$ is not divisible by $Z^2$. By the above picture, $\Pi \cap S$ is a set of 3 lines, and according to whether they are concurrent, I can arrange the coordinates so that

*either* (i) $\ell$: (T = 0), $\ell_1$: (X = 0), $\ell_2$: (Y = 0),

*or* (ii) $\ell$: (T = 0), $\ell_1$: (X = 0), $\ell_2$: (X = T).

Hence, in case (i), f = 2XYT + Zg, with g quadratic, and in terms of the expression (∗), this means that B = T, and Z | A, C, D, E, F. Therefore, modulo terms divisible by $Z^2$,

$$\Delta \equiv -T^2F \pmod{Z^2}.$$

In addition, the point P = (0,0,0,1) ∈ S, and non-singularity at P means that F must contain the term $ZT^2$ with non-zero coefficient. In particular, $Z^2$ does not divide F. Therefore (Z = 0) is a simple root of Δ.

Case (ii) is a similar calculation (see Ex. 7.1).

**(7.4) Corollary.** (a) There exist two disjoint lines $\ell, m \subset S$.

(b) S is rational (that is, birational to $\mathbb{P}^2$, see (5.9)).

**Proof.** (a) By (7.3), (ii), just take $\ell_1$ and $\ell_2$.

(b) Consider two disjoint lines $\ell, m \subset S$, and define rational maps

$$\varphi: S \dashrightarrow \ell \times m \quad \text{and} \quad \psi: \ell \times m \dashrightarrow S$$

as follows. If $P \in \mathbb{P}^3 \setminus (\ell \cup m)$ then there exists a unique line n through P which meets both $\ell$ and m:

$$P \in n, \text{ and } \ell \cap n \neq \emptyset, \ m \cap n \neq \emptyset.$$

Set $\Phi(P) = (\ell \cap n, \ m \cap n) \in \ell \times m$. This defines a morphism

$$\Phi: \mathbb{P}^3 \setminus (\ell \cup m) \to \ell \times m,$$

whose fibre above $(Q, R) \in \ell \times m$ is the line QR of $\mathbb{P}^3$. Define $\varphi: S \dashrightarrow \ell \times m$ as the restriction to S of Φ.

Conversely, for $(Q, R) \in \ell \times m$, let n be the line n = QR in $\mathbb{P}^3$. By (7.3), there are only finitely many lines of S meeting $\ell$, so that for almost all values of (Q, R), n intersects S in 3 points {P, Q, R}, of which Q and R are the given points on $\ell$ and m. Thus define $\psi: \ell \times m \dashrightarrow S$ by $(Q, R) \mapsto P$; then ψ is a rational map, since the ratios of coordinates of P are rational functions of those of Q, R.

Obviously φ and ψ are mutual inverses. Q.E.D.

**(7.5)** I want to find all the lines of S in terms of the configuration given by

Proposition 7.3 of a line $\ell$ and 5 disjoint pairs $(\ell_i, \ell_i')$. Any other line $n \subset S$ must meet exactly one of $\ell_i$ and $\ell_i'$ for $i = 1,.. 5$: this is because in $\mathbb{P}^3$, $n$ meets the plane $\Pi_i$, and $\Pi_i \cap S = \ell \cup \ell_i \cup \ell_i'$; also, $n$ cannot meet both $\ell_i$ and $\ell_i'$, since this would contradict (7.1, (a)). The key to sorting out the remaining lines is the following lemma, which tells us that $n$ is uniquely determined by which of $\ell_i$ and $\ell_i'$ it meets. Let me say that a line $n$ is a *transversal* of a line $\ell$ if $\ell \cap n \neq \emptyset$.

**Lemma.** If $\ell_1,.. \ell_4 \subset \mathbb{P}^3$ are disjoint lines then

*either* all 4 lines $\ell_i$ lie on a smooth quadric $\ell_1,.. \ell_4 \subset Q \subset \mathbb{P}^3$; and then they have an infinite number of common transversals;

*or* the 4 lines $\ell_i$ do not lie on any quadric $\ell_1,.. \ell_4 \not\subset Q$; and then they have either 1 or 2 common transversals.

**Proof.** There exists a smooth quadric $Q \supset \ell_1,.. \ell_3$: several proofs of this are possible (see Ex. 7.2).

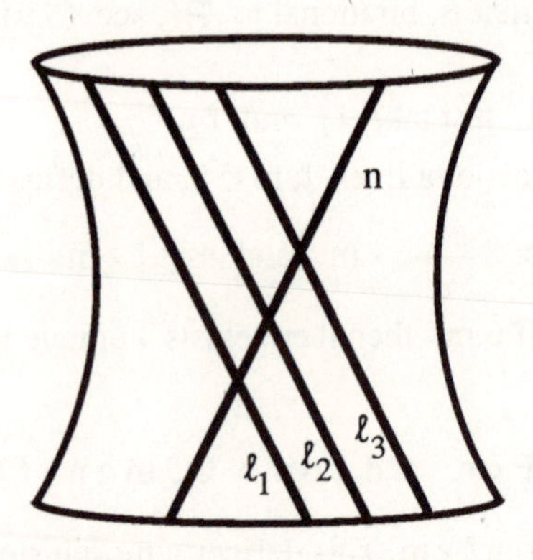


Then in some choice of coordinates, Q: (XT - YZ), and Q has two families of lines, or generators: any transversal of $\ell_1,.. \ell_3$ must lie in Q, since it has 3 points in Q. Now if $\ell_4 \not\subset Q$, then $\ell_4 \cap Q = \{1 \text{ or } 2 \text{ points}\}$, and the generators of the other family through these points are all the common transversals of $\ell_1,.. \ell_4$. Q.E.D.

**(7.6) The 27 lines.** Let $\ell$ and $m$ be two disjoint lines of S; as already observed, m meets exactly one out of each of the 5 pairs $(\ell_i, \ell_i')$ of lines meeting $\ell$. By renumbering the pairs, I assume that m meets $\ell_i$ for $i = 1,.. 5$. Introduce the following notation for the lines meeting $\ell$ or m:

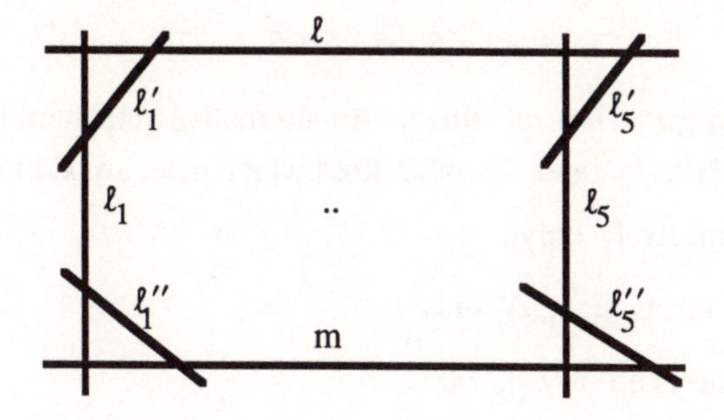


thus the 5 pairs of lines meeting m are $(\ell_i, \ell_i'')$ for $i = 1,.. 5$. By (7.3, (ii)) applied to m, for $i \neq j$, the line $\ell_i''$ does not meet $\ell_j$. On the other hand, every line of S must meet one of $\ell$, $\ell_j$ or $\ell_j'$, hence $\ell_i''$ meets $\ell_j'$ for $i \neq j$.

**Claim.** (I) If $n \subset S$ is any line other than these 12, then n meets exactly 3 out of the 5 lines $\ell_1,.. \ell_5$;

(II) conversely, given any choice of 3 elements $\{i, j, k\} \subset \{1, 2, 3, 4, 5\}$, there is a unique line $\ell_{ijk} \subset S$ meeting $\ell_i, \ell_j, \ell_k$.

**Proof.** (I) Given four disjoint lines of S, it is clear that they do not all lie on a non-singular quadric Q, since otherwise $Q \subset S$, contradicting the irreducibility of S. If n meets $\geq 4$ of the $\ell_i$, then by Lemma 7.5, $n = \ell$ or m, which is a contradiction.

If n meets $\leq 2$ of the $\ell_i$, then it must meet $\geq 3$ of the $\ell_i'$; say n meets $\ell_1$, $\ell_3'$, $\ell_4'$, $\ell_5'$ (and either $\ell_2$ or $\ell_2'$ as it thinks fit). But these 4 lines have $\ell$ and $\ell_1''$ as common transversals, by what was said above, so that again by Lemma 7.5, $n = \ell$ or $\ell_1''$. This is the same contradiction.

(II) There are 10 lines meeting $\ell_1$ by (7.3), of which so far only 4 have been accounted for (namely, $\ell$, $\ell_1'$, m and $\ell_1''$). The six other lines must meet exactly 2 out of the 4 remaining lines $\ell_2,.. \ell_5$, and there are exactly $6 = \binom{4}{2}$ possible choices; so they must all occur. Q.E.D.

This gives the lines of S as being

$$\{\ell, m, \ell_i, \ell_i', \ell_i'', \ell_{ijk}\},$$

and the number of them is

$$1 + 1 + 5 + 5 + 5 + 5 = 27.$$

**(7.7) The configuration of lines.** An alternative statement is that the lines of S are $\ell, \ell_1,.. \ell_5, \ell_1',.. \ell_5'$, and 16 other lines which meet an odd number of $\ell_1,.. \ell_5$:

$\ell_i''$ meets $\ell_i$ only

$\ell_{ijk}$ meets $\ell_i, \ell_j, \ell_k$ only

m meets all of $\ell_1,.. \ell_5$.

In the notation I have introduced, it is easy to see that the incidence relation between the 27 lines of S are as follows:

$\ell$ meets $\ell_1,.. \ell_5, \ell_1',.. \ell_5'$;

$\ell_1$ meets $\ell$, m, $\ell_1', \ell_1''$, and $\ell_{1jk}$ for 6 choices of $\{j, k\} \subset \{2,3,4,5\}$;

$\ell_1'$ meets $\ell, \ell_1, \ell_j''$ (for 4 choices of $j \neq 1$),

and $\ell_{ijk}$ (for 4 choices of $\{i, j, k\} \subset \{2,3,4,5\}$);

$\ell_1''$ meets m, $\ell_1, \ell_j'$ (for 4 choices of $j \neq 1$),

and $\ell_{ijk}$ (for 4 choices of $\{i, j, k\} \subset \{2,3,4,5\}$);

$\ell_{123}$ meets $\ell_1, \ell_2, \ell_3, \ell_{145}, \ell_{245}, \ell_{345}, \ell_4', \ell_5', \ell_4'', \ell_5''$.

This combinatorial configuration has many different representations, some of them much more symmetric than that given here; see for example [Semple and Roth, pp.122–128 and 151–152].

## Exercises to §7.

**7.1.** Prove case (ii) of the claim in Proposition 7.3. (Hint: as in the given proof of case (i), $f = X(X - T)T + Zg$, so that $A = T$, $D = -(1/2)T^2 + Z \cdot \ell$, where $\ell$ is linear, so that $Z | B, C, E, F$, and Z does not divide D; also, the non-singularity of S at (0,1,0,1) implies that $C = cZ$, with $c \neq 0$. Now calculate the determinant modulo $Z^2$).

**7.2.** Prove that given 3 disjoint lines $\ell_1,.. \ell_3 \subset \mathbb{P}^3$, there exists a non-singular quadric $Q \supset \ell_1,.. \ell_3$. (Hint: take 3 points $P_i, P_i', P_i'' \in \ell_i$ on each of them, and show as in (1.7) or (2.5) that there is at least one quadric Q through them; it follows that each $\ell_i \subset Q$. Now show that Q can't be singular: for example, what happens if Q is a pair of planes?)

**7.3.** Prove that if $S : (f = 0) \subset \mathbb{P}^3$ is a non-singular cubic surface and $\ell$ a line such that $f_{|\ell}$ has a triple zero at $P$ then the tangent plane $T_PS$ intersects $S$ in a cuspidal cubic. Deduce that in suitable coordinates,

$$f = X^2Z_0 - Y_0{}^3 + gT.$$

**7.4.** (i) Prove that if $P \in S$ is a singular point of a cubic surface then there is at least one line $\ell \subset S$ through $P$ (and 'in general' 6).

(ii) If $X \subset \mathbb{P}^4$ is a non-singular cubic hypersurface (a cubic 3-fold) and $P \in X$ then there is at least one line $\ell \subset X$ through $P$ (and 'in general' 6). (Hint: write down the equation of $X$ in coordinates with $P = (1, 0,.. 0)$.)

**7.5.** Prove that the rational map $\varphi: S \dashrightarrow \ell \times m$ of (7.3, (b)) is in fact a morphism; prove that it contracts 5 lines of $S$ to points.

**7.6.** Find all 27 lines of the diagonal (or 'Fermat') cubic surface

$$S: (X^3 + Y^3 + Z^3 + T^3 = 0) \subset \mathbb{P}^3$$

in terms of planes such as $(X = \rho Y)$, where $\rho^3 = 1$.

**7.7.** Let $S \subset \mathbb{P}^3$ be the cubic surface given by $S: (f = 0)$, where

$$f(X,Y,Z,T) = ZX^2 + TY^2 + (Z - d^2T)(Z - e^2T)(Z - f^2T),$$

with $d, e, f$ distinct non-zero elements of $k$, and $L \subset S$ the line given by $Z = T = 0$. By considering as in (7.3) the variable plane through $L$, write down the equations of the 10 lines of $S$ meeting $L$.

**7.8** (suggested by R.Casdagli). Consider the topology of the cubic surface $S_{(0)} \subset \mathbb{R}^3$ given in affine coordinates by

$$x^2 + y^2 + z^2 - 2xyz = 1 + \lambda,$$

where $\lambda > 0$ is a constant. By writing the equation as

$$(x - yz)^2 = y^2z^2 - y^2 - z^2 + 1 + \lambda,$$

show that $S_{(0)}$ has 4 tubes going off to infinity. On the other hand, the corresponding projective surface $S \subset \mathbb{P}^3_{\mathbb{R}}$ meets infinity in 3 lines $XYZ = 0$. Show that the eight lines of $S_{(0)}$ which meet $(Z = 0)$ asymptotically are given by

$$z = \pm 1, \quad x \pm y = \pm\sqrt{(1 + v)}.$$

Represent the surface $S_{(0)}$ in $\mathbb{R}^3$ and its 24 lines by computer graphics, or by making a plaster model.

# §8. Final comments

This final section is not for examination, but some of the topics may nevertheless be of interest to the student.

## History and sociology of the modern subject.

**(8.1) Introduction.** Algebraic geometry has over the last thirty years or so enjoyed a position in math similar to that of math in the world at large, being respected and feared much more than understood. At the same time, the 'service' questions I am regularly asked by British colleagues or by Warwick graduate students are generally of an elementary kind: as a rule, they are either covered in this book or in [Atiyah and Macdonald]. What follows is a view of the recent development of the subject, attempting to explain this paradox. I make no pretence at objectivity.

**(8.2) Prehistory.** Algebraic geometry developed in the 19th century from several different sources. Firstly, the geometrical tradition itself: projective geometry (and descriptive geometry, of great interest to the military at the time of Napoleon), the study of curves and surfaces for their own sake, configuration geometry; then complex function theory, the view of a compact Riemann surface as an algebraic curve, and the purely algebraic reconstruction of it from its function field. On top of this, the deep analogy between algebraic curves and the ring of integers of a number field, and the need for a language in algebra and geometry for invariant theory, which played an important role in the development of abstract algebra at the start of the 20th century.

The first decades of the 20th century saw a deep division. On the one hand, the geometric tradition of studying curves and surfaces, as pursued notably by the brilliant Italian school; alongside its own quite considerable achievements, this played a substantial motivating role in the development of topology and differential geometry, but became increasingly dependent on arguments 'by geometric intuition' that even the *Maestri* were unable to sustain rigorously. On the other hand, the newly emerging forces of commutative algebra were laying foundations and providing techniques of proof. An example of the difference between the two approaches was the argument between Chow and van der Waerden, who established rigorously the existence of an algebraic variety parametrising space curves of given degree and genus, and Severi, who had been making creative use of such parameter spaces all his working life, and who in his declining years bitterly resented the intrusion of algebraists (non-Italians at that!) into his field, and most especially the implicit suggestion that the work of his own school lacked rigour.

**(8.3) Rigour, the first wave.** Following the introduction of abstract algebra by Hilbert and Emmy Noether, rigorous foundations for algebraic geometry were laid in the 1920s and 1930s by van der Waerden, Zariski and Weil (van der Waerden's contribution is often supressed, apparently because a number of mathematicians of the immediate post-war period, including some of the leading algebraic geometers, considered him a Nazi collaborator).

A central plank of their program was to make algebraic geometry work over an arbitrary field. In this connection, a key foundational difficulty is that you can't just define a variety to be a point set: if you start life with a variety $V \subset \mathbb{A}^n_k$ over a given field $k$ then $V$ is not just a subset of $k^n$; you must also allow $K$-valued points of $V$ for field extensions $k \subset K$ (see (8.13, (c)) for a discussion). This is one reason for the notation $\mathbb{A}^n_k$, to mean the $k$-valued points of a variety $\mathbb{A}^n$ that one would like to think of as existing independently of the specified field $k$.

The necessity of allowing the ground field to change throughout the argument added enormously to the technical and conceptual difficulties (to say nothing of the notation). However, by around 1950, Weil's system of foundations was accepted as the norm, to the extent that traditional geometers (such as Hodge and Pedoe) felt compelled to base their books on it, much to the detriment, I believe, of their readability.

**(8.4) The Grothendieck era.** From around 1955 to 1970, algebraic geometry was dominated by Paris mathematicians, first Serre then more especially Grothendieck and his school. It is important not to underestimate the influence of Grothendieck's approach, especially now that it has to some extent gone out of fashion. This was a period in which tremendous conceptual and technical advances were made, and thanks to the systematic use of the notion of scheme (more general than a variety, see (8.12–14) below), algebraic geometry was able to absorb practically all the advances made in topology, homological algebra, number theory, etc., and even to play a dominant role in their development. Grothendieck himself retired from the scene around 1970 in his early forties, which must be counted a tragic waste (he initially left the IHES in a protest over military funding of science). As a practising algebraic geometer, one is keenly aware of the large blocks of powerful machinery developed during this period, many of which still remain to be written up in an approachable way.

On the other hand, the Grothendieck personality cult had serious side effects: many people who had devoted a large part of their lives to mastering Weil foundations suffered rejection and humiliation, and to my knowledge only one or two have adapted to the new language; a whole generation of students (mainly French) got themselves brain-washed into the foolish belief that a problem that can't be dressed up in high-powered abstract formalism is unworthy of study, and were thus excluded from the mathematician's natural development of starting with a small problem he can handle and exploring outwards from there. (I actually know of a thesis on the arithmetic of cubic surfaces that was initially not considered because 'the natural context for the construction is over a general locally Noetherian ringed topos'. This is not a joke.) Many students of the time could apparently not think of any higher ambition than *étudier les EGAs* . The study of category theory for its own sake (surely one of the most sterile of all intellectual pursuits) also dates from this time; Grothendieck himself can't necessarily be blamed for this, since his own use of categories was very successful in solving problems.

The fashion has since swung the other way. At a recent conference in France I commented on the change in attitude, and got back the sarcastic answer 'but the twisted cubic is a very good example of a pro-representable functor'. I understand that some of the mathematicians now involved in administering French research money are individuals who suffered during this period of intellectual terrorism, and that applications for CNRS research projects are in consequence regularly dressed up to minimise their connection with algebraic geometry.

Apart from a very small number of his own students who were able to take the pace and survive, the people who got most lasting benefit from Grothendieck's ideas, and who have propagated them most usefully, were influenced at a distance: the Harvard school (through Zariski, Mumford and M. Artin), the Moscow school of Shafarevich, perhaps also the Japanese school of commutative algebraists.

**(8.5) The big bang.** History did not end in the early 1970s, nor has algebraic geometry been less subject to swings of fashion since then. During the 1970s, although a few big schools had their own special interests (Mumford and compactification of moduli spaces, Griffiths' schools of Hodge theory and algebraic curves, Deligne and 'weights' in the cohomology of varieties, Shafarevich and K3 surfaces, Iitaka and his followers in the classification of higher-dimensional varieties, and so on), it seems to me we all basically believed we were studying the same subject, and that algebraic geometry remained a monolithic block (and was in fact colonising adjacent areas of math). Perhaps the presence of just one or two experts who could handle the whole range of the subject made this possible.

By the mid-1980s, this had changed, and algebraic geometry seems at present to be split up into a dozen or more schools having quite limited interaction: curves and Abelian varieties, algebraic surfaces and Donaldson theory, 3-folds and classification in higher dimensions, K-theory and algebraic cycles, intersection theory and enumerative geometry, general cohomology theories, Hodge theory, characteristic p, arithmetic algebraic geometry, singularity theory, differential equations of math physics, string theory, applications of computer algebra, etc.

## Additional footnotes and high-brow comments.

This section mixes elementary and advanced topics; since it is partly a 'word to the wise' for university teachers using this as a textbook, or to guide advanced students into the pitfalls of the subject, some of the material may seem obscure.

**(8.6) Choice of topics.** The topics and examples treated in this book have been chosen partly pragmatically on the basis of small degree and ease of computation. However, they also hint at the 'classification of varieties': the material on conics applies in a sense to every rational curve, and cubic surfaces are the most essential examples of del Pezzo rational surfaces. Cubic curves with their group law are examples of Abelian varieties; the fact (2.2) that a non-singular cubic is not rational is the very first step in classification. The intersection of two plane conics in (1.12–14) and the intersection of two quadrics of $\mathbb{P}^3{}_k$ referred to in Ex. 5.6 could also be fitted into a similar pattern, with the intersection of two quadrics in $\mathbb{P}^4{}_k$ providing another class of del Pezzo surfaces, and the family of lines on the intersection of two quadrics in $\mathbb{P}^5{}_k$ providing 2-dimensional Abelian varieties.

The genus of a curve, and the division into 3 cases tabulated on p.46 is classification in a nutshell. I would have liked to include more material on the genus of a curve, in particular how to calculate it in terms of topological Euler characteristic or of intersection numbers in algebraic geometry, essential five-finger exercises for young geometers. However, this would comfortably occupy a separate undergraduate lecture course, as would the complex analytic theory of elliptic curves.

**(8.7) Computation versus theory.** Another point to make concerning the approach in these notes is that quite a lot of emphasis is given to cases that can be handled by explicit calculations. When general theory proves the existence of some construction, then doing it in terms of explicit coordinate expressions is a useful exercise that helps one to keep a grip on reality, and this is appropriate for an undergraduate textbook. This should not however be allowed to obscure the fact that the theory is really designed to handle the complicated cases, when explicit computations will often not tell us anything.

**(8.8) $\mathbb{R}$ versus $\mathbb{C}$.** The reader with real interests may be disappointed that the treatment over $\mathbb{R}$ in §§1-2 gave way in §3 to considerations over an arbitrary field k, promptly assumed to be algebraically closed. I advise this class of reader to persevere; there are plenty of relations between real and complex geometry, including some that will come as a surprise. Asking about the real points of a real variety is a very hard question, and something of a minority interest in algebraic geometry; in any case, knowing all about its complex points will usually be an essential prerequisite. Another direct relation between geometry over $\mathbb{R}$ and $\mathbb{C}$ is that an n-dimensional non-singular complex variety is a 2n-dimensional real manifold - for example, algebraic surfaces are a principal source of constructions of smooth 4-manifolds.

As well as these fairly obvious relations, there are more subtle ones, for example: (a) singularities of plane curves $C \subset \mathbb{C}^2$ give rise to knots in $S^3$ by intersecting with the boundary of a small ball; and (b) the Penrose twistor construction views a 4-manifold (with a special kind of Riemannian metric) as the set of real-valued points of a 4-dimensional complex variety that parametrises rational curves on a complex 3-dimensional variety (thus the real 4-sphere $S^4$ we live in can be identified as the real locus in the complex Grassmannian Gr(2, 4) of lines in $\mathbb{P}^3_{\mathbb{C}}$).

**(8.9) Regular functions and sheaves.** The reader who has properly grasped the notion of rational function $f \in k(X)$ on a variety X and of regularity of f at $P \in X$ ((4.7) and (5.4)) already has a pretty good intuitive idea of the structure sheaf $\mathcal{O}_X$. For an open set $U \subset X$, the set of regular functions $U \to k$

$$\mathcal{O}_X(U) = \{ f \in k(X) \mid f \text{ is regular } \forall P \in U \} = \bigcap_{P \in U} \mathcal{O}_{X,P}$$

is a subring of the field k(X). The sheaf $\mathcal{O}_X$ is just the family of rings $\mathcal{O}_X(U)$ as U runs through the opens of X. Clearly, any element of the local ring $\mathcal{O}_{X,P}$ (see (4.7) and (5.4) for the definition) is regular in some neighbourhood U of P, so that $\mathcal{O}_{X,P} = \bigcup_{P \in U} \mathcal{O}_X(U)$. There's no more to it than that; there's a fixed pool of rational sections k(X), and sections of the sheaf over an open U are just rational sections with a regularity condition at every $P \in U$.

This language is adequate to describe any torsion free sheaf on an irreducible variety with the Zariski topology. Of course, you need the full definition of sheaves if X is reducible, or if you want to handle more complicated sheaves, or to use the complex topology.

**(8.10) Globally defined regular functions.** If X is a projective variety then the only rational functions $f \in k(X)$ that are regular at every $P \in X$ are the constants. This is a general property of projective varieties, analogous to Liouville's theorem in functions of one complex variable; for a variety over $\mathbb{C}$ it comes from compactness and the maximum modulus principle ($X \subset \mathbb{P}^n_{\mathbb{C}}$ is compact in the complex topology, so the modulus of a global holomorphic function on $X$ must take a maximum), but in algebraic geometry it is surprisingly hard to prove from scratch (see for example [Hartshorne, I.3.4]; it is essentially a finiteness result, related to the finite-dimensionality of coherent cohomology groups).

**(8.11) The surprising sufficiency of projective algebraic geometry.** Weil's abstract definition of a variety (affine algebraic sets glued together along isomorphic open sets) was referred to briefly in (0.4), and is quite easy to handle in terms of sheaves. Given this, the idea of working only with varieties embedded in a fixed ambient space $\mathbb{P}^N_k$ seems at first sight unduly restrictive. I want to describe briefly the modern point of view on this question.

**(a) Polarisation and positivity.** Firstly, varieties are usually considered up to isomorphism, so saying a variety $X$ is *projective* means that $X$ can be embedded in some $\mathbb{P}^N$, that is, is isomorphic to a closed subvariety $X \subset \mathbb{P}^N$ as in (5.1-7). *Quasiprojective* means isomorphic to a locally closed subvariety of $\mathbb{P}^N$, so an open dense subset of a projective variety; projectivity includes the property of *completeness*, that $X$ cannot be embedded as a dense open set of any bigger variety.

The choice of an actual embedding $X \hookrightarrow \mathbb{P}^N$ (or of a very ample line bundle $\mathcal{O}_X(1)$ whose sections will be the homogeneous coordinates of $\mathbb{P}^N$) is often called a *polarisation*, and we write $(X, \mathcal{O}_X(1))$ to indicate that the choice has been made. In addition to completeness, a projective variety $X \subset \mathbb{P}^N$ satisfies a key condition of 'positive degree': if $V \subset X$ is a $k$-dimensional subvariety then $V$ intersects a general linear subspace $\mathbb{P}^{N-k}$ in a positive finite number of points. Conversely, the Kleiman criterion says that some multiple of a line bundle on a complete variety $X$ can be used to provide a projective embedding of $X$ if its degree on every curve $C \subset X$ is consistently greater than zero (that is, $\geq \varepsilon \cdot$(any reasonable measure of $C$)). This kind of positivity relates closely to the choice of a Kähler metric on a complex manifold (a Riemannian metric with the right kind of compatibility with the complex structure). So we understand projectivity as a kind of 'positive definiteness'.

**(b) Sufficiency.** The surprising thing is the many problems of algebraic geometry having answers within the framework of projective varieties. The construction of Chow varieties mentioned in (8.2) is one such example; another is Mumford's work of the 1960s, in which he constructed Picard varieties and many moduli spaces as quasiprojective varieties (schemes). Mori theory (responsible for important conceptual advances in classification of varieties related to rationality, see [Kollár]) is the most recent example; here the ideas and techniques are inescapably projective in nature.

**(c) Insufficiency of abstract varieties.** Curves and non-singular surfaces are automatically quasiprojective; but abstract varieties that are not quasiprojective do exist (singular surfaces, or non-singular varieties of dimension $\geq 3$). However, if you feel the need for these constructions, you will almost certainly also want Moishezon varieties (M. Artin's algebraic spaces), objects of algebraic geometry more general than abstract varieties, obtained by a somewhat more liberal interpretations of 'glueing local pieces'.

Theorems on abstract varieties are often proved by a reduction to the quasiprojective case, so whether the quasiprojective proof or the detail of the reduction process is more useful, interesting, essential or likely-to-lead-to-cheap-publishable-work will depend on the particular problem and the individual student's interests and employment situation. It has recently been proved that a non-singular abstract variety or Moishezon variety that is not quasiprojective necessarily contains a rational curve; however, the proof (due to J. Kollár) is Mori-theoretic, so hardcore projective algebraic geometry.

**(8.12) Affine varieties and schemes.** The coordinate ring $k[V]$ of an affine algebraic variety $V$ over an algebraically closed field $k$ (Definition 4.1) satisfies two conditions: (i) it is a finitely generated $k$-algebra; and (ii) it is an integral domain. A ring satisfying these two conditions is obviously of the form $k[V]$ for some variety $V$, and is called a *geometric ring* (or *geometric $k$-algebra*).

There are two key theoretical results in Chapter II; one of these is Theorem 4.4, which states precisely that $V \mapsto k[V] = A$ is an equivalence of categories between affine algebraic varieties and the opposite of the category of geometric $k$-algebras (although I censored out all mention of categories as unsuitable for younger readers). The other is the Nullstellensatz (3.10), that prime ideals of $k[V]$ are in bijection with irreducible subvarieties of $V$; the points of $V$ are in bijection with maximal ideals.

Taken together, these results identify affine varieties $V$ with the affine schemes corresponding to geometric rings (compare also Definition 4.6).

The *prime spectrum* Spec $A$ is defined for an arbitrary ring (commutative with a 1) as the set of prime ideals of $A$. It has a Zariski topology and a structure sheaf; this is the *affine scheme* corresponding to $A$ (for details see [Mumford, Introduction, or Hartshorne, Ch. II]). There are several quite distinct ways in which affine schemes are more general than affine varieties; each of these is important, and I run through them briefly in (8.14).

It's important to understand that for a geometric ring $A = k[V]$, the prime spectrum Spec A contains exactly the same information as the variety V, and no more. The NSS tells us there's a plentiful supply of maximal ideals ($m_v$ for points $v \in V$), and every other prime P of A is the intersection of maximal ideals over the points of an irreducible subvariety $Y \subset V$:

$$P = I(Y) = \bigcap_{v \in Y} m_v;$$

It's useful and (roughly speaking, at least) permissible to ignore the distinction between varieties and schemes, writing V = Spec A, v for $m_v$, and imagining the prime P = I(Y) ('generic point') as a kind of laundry mark stitched everywhere dense into the fabric of the subvariety Y.

**(8.13) What's the point?** A majority of students will never need to know any more about scheme theory than what is contained in (8.9) and (8.12), beyond the warning that the expression *generic point* is used in several technical senses, often meaning something quite different from *sufficiently general* point.

This section is intended for the reader who faces the task of working with the modern literature, and offers some comments on the various notions of point in scheme theory, potentially a major stumbling-block for beginners.

**(a) Scheme-theoretic points of a variety.** Suppose that k is a field (maybe not algebraically closed), $I \subset k[X_1,.. X_n]$ an ideal and $A = k[X_1,.. X_n]/I$; write $V = V(I) \subset K^n$ where $k \subset K$ is a chosen algebraic closure. The points of Spec A are only a bit more complicated than for a geometric ring in (8.12). By an obvious extension of the NSS, a maximal ideal of A is determined by a point $v = (a_1,.. a_n) \in V \subset K^n$, that is, it's of the form

$$m_v = \{f \in A \mid f(P) = 0\} = (x_1-a_1,.. x_n-a_n) \cap A.$$

It's easy to see that different points $v \in V \subset K^n$ give rise to the same maximal ideal $m_v$ of A if and only if they are conjugate over k in the sense of Galois theory (since A consists of polynomials with coefficients in k). So the maximal spectrum Specm A is just V 'up to conjugacy' (the orbit space of Gal K/k on V). Every other prime P of A corresponds as in (8.12) to an irreducible subvariety $Y = V(P) \subset V$ (up to conjugacy over k); $P \in$ Spec A is the scheme-theoretic *generic point* of Y, and is again to be thought of as a laundry mark on Y. The Zariski topology of Spec A is fixed up so that P is everywhere dense in Y. The maximal ideals of A are called *closed points* to distinguish them. If C : (f = 0) $\subset \mathbb{A}^2_{\mathbb{C}}$ is an irreducible curve, it has just one scheme-theoretic generic point, corresponding to the ideal (0) of $\mathbb{C}[X, Y]/(f)$, whereas a surface S will have one generic point in each irreducible curve $C \subset S$ as well as its own generic point dense in S.

Scheme-theoretic points are crucial in writing down the definition of Spec A as a set with a topology and a sheaf of rings (and are also important in commutative algebra, and in the treatment in algebraic geometry of notions like the neighbourhood of a generic point of an irreducible subvariety, see (8.14, (i))); however, points of $V \subset K^n$ with values in the algebraically closure $k \subset K$ correspond more to the geometrical idea of a point, and are called *geometric points*. This is similar to the way that the Zariski topology of a variety $V$ serves more as a vehicle for the structure sheaf $\mathcal{O}_V$ than as a geometric object in its own right.

**(b) Field-valued points in scheme theory.** If $P$ is a prime ideal of $A$ (so $P \in \operatorname{Spec} A$ a point) the residue field at $P$ is the field of fractions of the integral domain $A/P$, written $k(P)$; it is an algebraic extension of the ground field $k$ if and only if $P$ is maximal. A point of $V$ with coefficients in a field extension $k \subset L$ (a point $(a_1,\dots a_n) \in V(I) \subset L^n$) clearly corresponds to a homomorphism $A \to L$ (given by $X_i \mapsto a_i$), with kernel a prime ideal $P$ of $A$, or equivalently, to an embedding $k(P) \hookrightarrow L$. If $P = m_v$ is a maximal ideal, and $L = K$ is the algebraic closure of $k$, it is the choice of the embedding $A/m_v = k(v) \hookrightarrow K$ that determines the coordinates of the corresponding point of $V \subset K^n$, or in other words distinguishes this point from its Galois conjugates. These are the geometric points of $V$.

For any extension $k \subset L$, the $k$–algebra homomorphism $A \to L$ corresponding to an $L$–valued point of $V$ can be dressed up to seem more reasonable. Recall first that a variety is more than a point set; even if it's only a single point, you have to say what field it's defined over. So

$$\operatorname{Spec} L = \overset{L}{\cdot} = \mathrm{pt}_L$$

is the variety consisting of a single point defined over $L$. By the equivalence of categories (4.4), a morphism $\operatorname{Spec} L \to V$ (the inclusion of a point defined over $L$) should be the same thing as a $k$-algebra homomorphism $A = k[V] \to L = k[\mathrm{pt}_L]$.

To summarise the relation betwen scheme-theoretic points and field-valued points: a point $P \in \operatorname{Spec} A = V$ is a prime ideal of $A$, so corresponds to the quotient homomorphism $A \to A/P \subset \operatorname{Quot}(A/P) = k(P)$ to a field. If $L$ is any field, a $L$–valued point of $V$ is a homomorphism $A \to L$; a scheme-theoretic point $P$ corresponds in a tautological way to a field-valued point, but with the field $k(P)$ varying with $P$. If $K$ is the algebraic closure of $k$ then $K$-valued points of $V \subset K^n$ are just geometric points; a $K$-valued point $v$ sits at a closed scheme-theoretic point $m_v$, with a specified inclusion $A/m_v = k(v) \hookrightarrow K$.

**(c) Generic points in Weil foundations.** I mentioned in (8.3) the peculiarity of points in Weil foundations: a variety $V$ defined over a field $k$ is allowed to have $L$-valued points for any field extension $k \subset L$. This clearly derives from number theory, but it also has consequences in geometry. For example, if $C$

is the circle $x^2 + y^2 = 1$ defined over $k = \mathbb{Q}$, then

$$P_\pi = (2\pi/(\pi^2 + 1), (\pi^2 - 1)/(\pi^2 + 1))$$

is allowed as a $\mathbb{C}$–valued point of C. Since $\pi$ is transcendental over $\mathbb{Q}$, any polynomial $f \in \mathbb{Q}[x, y]$ vanishing at $P_\pi$ is a multiple of $x^2 + y^2 - 1$; so $P_\pi$ is a $\mathbb{Q}$-*generic* point of C - it's not in any smaller subvariety of C defined over $\mathbb{Q}$. In other words, the conjugates of $P_\pi$ under Aut $\mathbb{C}$ (= "Gal ($\mathbb{C}/\mathbb{Q}$)") are dense in C. Since $P_\pi$ is $\mathbb{Q}$–generic, if you prove a statement only involving polynomials over $\mathbb{Q}$ about $P_\pi$, the same statement will be true for every point of C.

In fact this idea is already covered by the notion of an L–valued point described in (b), and the geometric content of generic points can be seen most clearly in this language. For example, the field $\mathbb{Q}(\pi)$ is just the purely transcendental extension, so $\mathbb{Q}(\pi) \cong \mathbb{Q}(\lambda)$ and the morphism Spec $\mathbb{Q}(\lambda) \to C$ is the rational parametrisation of C discussed in (1.1): roughly, you're allowed to substitute any 'sufficiently general' value for the transcendental or unknown $\pi$. More generally, a finitely generated extension $k \subset L$ is the function field of a variety W over k; suppose that $\varphi$: Spec $L \to V$ = Spec A is a point corresponding to a k–algebra homomorphism $A \to L$, having kernel P. Then $\varphi$ extends to a rational map f: W - → V whose image is dense in the subvariety $Y = V(P) \subset V$, so $\varphi$ or $\varphi$(Spec L) is a field–valued generic point of Y.

**(d) Points as morphisms in scheme theory.** The discussion in (c) shows that an L–valued point of a variety V contains implicitly a rational map W - → V where W is a variety birational to Spec L (that is, L = k(W)); a geometer could think of this as a family of points parametrised by W.

More generally, for X a variety (or scheme) we are interested in, an S–valued point of X (where S is any scheme) can just be defined as a morphism $S \to X$. If $X = V(I) \subset \mathbb{A}^n_k$ is affine with coordinate ring k[X] and S = Spec A, then an S–valued point corresponds under (4.4) to a k–algebra homomorphism $k[X] \to A$, that is, to an n-tuple $(a_1,.. a_n) \in A$ satisfying $f(\underline{a}) = 0$ for all $f \in I$.

In a high-brow sense, this is the final apotheosis of the notion of a variety: if a point of a variety X is just a morphism, then X itself is just the functor

$$S \mapsto X(S) = \{\text{morphisms } S \to X\}$$

on the category of schemes. (The fuss I made about the notation $\mathbb{A}^n_k$ in the footnote on p.50 already reflect this.) Unlikely as it may seem, these metaphysical incantations are technically very useful, and varieties defined as functors are basic in the modern view of moduli spaces. Given a geometrical construction that can 'depend algebraically on parameters' (such as space curves of fixed degree and genus), you can ask to endow the set of all possible constructions with the structure of an algebraic variety. Even better, you could ask for a family of constructions over a parameter space that is 'universal', or 'contains all possible constructions'; the parameter variety of this universal family can usually be defined most directly as a functor (you still have to prove that the variety exists). For example the Chow variety referred to in (8.2) represents the functor

$$S \mapsto \{\text{families of curves parametrised by } S\}.$$

**(8.14) How schemes are more general than varieties.** I now discuss in isolation 3 ways in which affine scheme are more general than affine varieties; in cases of severe affliction, these complications may occur in combination with each other, with the global problems discussed in (8.11), or even in combination with new phenomena such as p-adic convergence or Arakelov Hermitian metrics. Considerations of space fortunately save me from having to say more on these fascinating topics.

(i) **Not restricted to finitely generated algebras.** Suppose $C \subset S$ is a curve on a non-singular affine surface (over $\mathbb{C}$, if you must). The ring

$$\mathcal{O}_{S,C} = \{f \in k(S) \mid f = g/h \text{ with } h \notin I_C\} \subset k(S)$$

is the *local ring* of $S$ at $C$; elements $f \in \mathcal{O}_{S,C}$ are regular on an open set of $S$ containing a dense open subset of $C$. Divisibility theory in this ring is very splendid, and relates to the geometrical idea of zeros and poles of a meromorphic function: $C$ is locally defined by a single equation $(y = 0)$ with $y \in I_C$ a local generator, and every nonzero element $f \in \mathcal{O}_{S,C}$ is of the form $f = y^n \cdot f_0$, where $n \in \mathbb{Z}$ and $f_0$ is an invertible element of $\mathcal{O}_{S,C}$. A ring with this property is called a *discrete valuation ring* (d.v.r.), in honour of the discrete valuation $f \mapsto n$, which counts the order of zero of $f$ along $C$ ($n < 0$ corresponds to poles); the element $y$ is called a *local parameter* of $\mathcal{O}_{S,C}$.

Now scheme theory allows us boldly to consider $\operatorname{Spec} \mathcal{O}_{S,C}$ as a geometric object, the topological space $(\cdot -)$ with only two points: a closed point, the maximal ideal $(y)$ (= the generic point of $C$) and a nonclosed point, the zero ideal $0$ (= the generic point of $S$). The advantage here is not so much technical: the easy commutative algebra of discrete valuation rings was of course used to prove results in algebraic geometry and complex function theory (for example, about ideals of functions, or about the local behaviour above $C$ of a branched cover $T \to S$ in terms of the field extension $k(S) \subset k(T)$) long before schemes were invented. More important, it gives us a precise geometric language, and a simple picture of the local algebra.

The above is just one example, related to localisation, or the idea of 'neighbourhood of a generic point of a subvariety', of benefits to ordinary geometry from taking Spec of a ring more general than a finitely generated algebra over a field; a similar example is thinking of the generic point $\operatorname{Spec} k(W)$ of a variety $W$

as the variety obtained as the intersection of all nonempty open sets of W (compare (8.14, (d))), like the grin remaining after the Cheshire cat's face has disappeared.

**(ii) Nilpotents.** The ring A can have nilpotent elements; for example $A = k[x, y]/(y^2 = 0)$ corresponds to the 'double line' $2\ell \subset \mathbb{A}^2_k$, to be thought of as an infinitesimal strip neighbourhood of the line. An element of A is of the form $f(x) + \varepsilon f_1(x)$ (with $\varepsilon^2 = 0$), so it looks like a Taylor series expansion of a polynomial about $\ell$ truncated to first order. If you practise hard several times a day, you should be able to visualise this as a function on the double line $2\ell$.

Nilpotents allow scheme theory to deal in Taylor series truncated to any order, so for example to deal with points of a variety by power series methods. They are crucial in the context of the moduli problems discussed at the end of (8.14, (d)): for example, they provide a precise language for handling first order infinitesimal deformations of a geometric construction (as a construction over the parameter space $\operatorname{Spec}(k[\varepsilon]/(\varepsilon^2 = 0)))$ and viewing these as tangent vectors to the universal parameter variety. They also open up a whole range of phenomena for which there was no classical analogue, for example relations between inseparable field extensions and Lie algebras of vector fields on varieties in characteristic p.

**(iii) No base field.** Let p be a prime number, and $\mathbb{Z}_{(p)} \subset \mathbb{Q}$ the subring of rationals with no p in the denominator; $\mathbb{Z}_{(p)}$ is another discrete valuation ring, with parameter p. It has a unique maximal ideal $0 \neq p\mathbb{Z}_{(p)}$, with residue field $\mathbb{Z}_{(p)}/p\mathbb{Z}_{(p)} \cong \mathbb{F}_p = \mathbb{Z}/(p)$. If $F \in \mathbb{Z}_{(p)}[X, Y]$, then it makes sense to consider the curve $C_{\mathbb{C}} : (F = 0) \subset \mathbb{A}^2_{\mathbb{C}}$, or alternatively to take the reduction f of F mod p, and to consider the curve $C_p: (f = 0) \subset \mathbb{A}^2_{\mathbb{F}_p}$. What kind of geometric object is it that contains both a curve over the complexes and a curve over a finite field? Whether you consider it to be truly geometric is a matter of opinion, but the scheme $\operatorname{Spec} \mathbb{Z}_{(p)}[X, Y]/(F)$ does exactly this.

Again, this is technically not a new idea: reducing a curve mod p has been practised since the 18th century, and Weil foundations contained a whole theory of 'specialisation' to deal with it. The advantage is a better conceptual picture of the curve $\operatorname{Spec} \mathbb{Z}_{(p)}[X, Y]/(F)$ over the d.v.r. $\mathbb{Z}_{(p)}$ as a geometric object fibred over $\operatorname{Spec}(\mathbb{Z}_{(p)})$ ('= (·–)'), with the two curve $C_{\mathbb{C}}$ and $C_p$ as generic and special fibre.

In the same way, for $F \in \mathbb{Z}[X, Y]$, the scheme $\operatorname{Spec} \mathbb{Z}[X, Y]/(F)$ is a geometric object containing for every prime p the curve $C_p: (f_p = 0) \subset \mathbb{A}^2_{\mathbb{F}_p}$ over $\mathbb{F}_p$, where $f_p$ is the reduction of F mod p, and at the same time the curve $C_{\mathbb{C}} : (F = 0) \subset \mathbb{A}^2_{\mathbb{C}}$, and is called an *arithmetic surface*; it contains quite a lot besides: in particular, for every point $c \in C_{\mathbb{C}}$ with algebraic numbers as coordinates, it contains a copy of $\operatorname{Spec} \mathbb{Q}[c]$, hence essentially all the information about the ring of integers of the number field $\mathbb{Q}(c)$ of definition of c.

However grotesquely implausible this object may seem at first sight (you can again get used to it if you practise), it is a key ingredient in modern number theory, and is the basic foundation on which the work of Arakelov and Faltings rests.

**(8.15) Proof of the existence of lines on a cubic surface.** Every adult algebraic geometer knows the traditional proof of (7.2) by dimension-counting (see for example [Beauville, Complex algebraic surfaces, p.50, or Mumford, Algebraic geometry I, Complex projective varieties, p.174]). I run through this before commenting on the difficulties.

The set of lines of $\mathbb{P}^3$ is parametrised by the 4-dimensional Grassmanian $\mathrm{Gr} = \mathrm{Gr}(2, 4)$, and cubic surfaces by the projective space $S = \mathbb{P}^N$ of cubic forms in $(X, Y, Z, T)$ (in fact $N = 19$). Write $Z \subset \mathrm{Gr} \times S$ for the incidence subvariety

$$Z = \{(\ell, X), \ell \in \mathrm{Gr}, X \in S \mid \ell \subset X\}.$$

Since cubic forms vanishing on a given line $\ell$ form a $\mathbb{P}^{N-4}$, it is easy to deduce from the first projection $Z \to \mathrm{Gr}$ that $Z$ is a rational $N$-dimensional variety. So the second projection $p\colon Z \to S$ is a morphism between two $N$-dimensional varieties, and therefore

(i) *either* the image $p(Z)$ is an $N$-dimensional variety in $S$, and so contains a dense open of $S$, *or* every fibre of $p$ has dimension $\geq 1$.

(ii) $Z$ is a projective variety, so that the image $p(Z)$ is closed in $S$.

Since cubic surfaces containing only finitely many lines do exist, the second possibility in (i) doesn't occur, so every sufficiently general cubic surface contains lines. Then (ii) ensures that $p(Z) = S$, and every cubic surface contains lines.

This argument seems to me to be unsuitable for an undergraduate course for two reasons: statement (i) assumes results about the dimension of fibres, which however intuitively acceptable (especially to students in the last week of a course) are hard to do rigorously; whereas (ii) is the theorem that a projective variety is complete, that again requires proof (by elimination theory, compactness, or a full-scale treatment of the valuative criterion for properness).

To the best of my knowledge, my proof in (7.2) is new; the knowledgeable reader will of course see its relation to the other traditional argument by vector bundles: the Grassmanian $\mathrm{Gr}(2, 4)$ has a tautological rank 2 vector bundle $E$ (consisting of linear forms on the lines of $\mathbb{P}^3$); restricting the equation $f$ of a cubic surface to every line $\ell \subset \mathbb{P}^3$ defines a section $s(f) \in S^3E$ of the 3rd symmetric power of $E$. Finally, every section of $S^3E$ must have a zero, either by ampleness of $E$ or by a Chern class argument (that also gives the magic number 27).

## Substitute for preface.

**(8.16) Acknowledgements and name-dropping.** It would be futile to try to list all the mathematicians who have contributed to my education. I owe a great debt to both my formal supervisors Pierre Deligne and Peter Swinnerton-Dyer (before he became a successful politician and media personality); I probably learned most from the books of David Mumford, and my understanding (such as it is) of the Grothendieck legacy derives largely from Mumford and Deligne. My view of the world, both as a mathematician and as a human being, has been strongly influenced by Andrei Tyurin.

My approach to what an undergraduate algebraic geometry should be is partly based on a course designed around 1970 by Swinnerton-Dyer for the Cambridge tripos, and taught in subsequent years by him and Barry Tennison; my book is in some ways a direct descendant of this, and some of the exercises have been taken over verbatim from Barry Tennison's example sheets. However, I have benefitted enormously from the freedom allowed under the Warwick course structure, especially the philosophy of teaching (explicitly stated by Christopher Zeeman) that research experience must serve as one's main guideline in deciding how and what to teach.

The 'winking torus' appearing in (2.14-) comes to me from Jim Eells, who informs me he learnt it from H. Hopf (and that it probably goes back to an older German tradition of mathematical art-work). I must thank Caroline Series, Frans Oort, Paul Cohn, John Jones, Ulf Persson, David Fowler, an anonymous referee and David Tranah from C.U.P. for helpful comments on the preprint version of this book, and apologise if on occasions I have either not been fully able to accommodate their suggestions, or preferred my own counsel.